T0182177

The Parasite Chronicles

Stillness in motion

Dawn glows, and waking sky forsakes
      The fading star's slow lingerings.
Silent, misty, by the lakes,
      A white crane spreads its wings.

The mighty pine dissolves in mist,
      Yielding, but rooted to stay.
Branches relaxed, yet force resist -
      Inspiring the Ultimate Way.

Half-moon floats o'er darkened eaves,
      Cat glides with silent paws.
Tranquil, this night receives
      Insights to the Seven Stars.

(By Ke Wuhe, a humble acolyte of Tai Chi
Master Chen Li-tsun)

Boo H. Kwa

The Parasite Chronicles

My Lifelong Odyssey Among the Parasites
that Cause Human Disease

 Springer

Boo H. Kwa
College of Public Health
University of South Florida
Tampa, Florida, USA

ISBN 978-3-030-09105-7   ISBN 978-3-319-74923-5   (eBook)
https://doi.org/10.1007/978-3-319-74923-5

Cover image credit: @toeytoey/Shutterstock

Printed on acid-free paper

This Springer imprint is published by the registered company Springer International Publishing AG part of
Springer Nature.
The registered company address is: Gewerbestrasse 11, 6330 Cham, Switzerland

To my parents whose sacrifices made this journey possible; to my wife whose loving companionship made this journey a joy; and to my daughters and grandchildren who made this journey worthwhile.

Foreword by B. Kim Lee Sim

I know why I cannot bring myself to eat raw fish. It is because of an incredibly charismatic and inspirational professor who delivered theatrical lectures, about intricate life cycles and the derivation of the complicated binomial nomenclature of slews of scientific names for weird organisms, peppered with delightfully ghastly tales about these very icky creatures and what they could do to their human hosts. Dr. Kwa's fabled parasitology course at the University of Malaya was the most popular course at our university. I remember clearly Dr. Kwa, with a crooked smile, dryly relating a yarn about how students were fooled by their professor into drinking pinworm eggs for the sake of good experimental science. So it seemed meant to be that I sought Dr. Kwa to be my mentor for PhD studies at the University of Malaya.

Dr. Kwa has been a guiding force in my life for nearly four decades. I am not alone. He has mentored, lectured, and educated generations of scientists over the course of his career on both sides of the globe. This is in itself already exceptional, but yet, no one tells a tale like Kwa.

The *Parasite Chronicles* takes us into a dimension less traveled. It is filled with stories, narrated with a wry, slyly evil smile, about the curious, complicated parasites that have been the focus of his scientific efforts. The stories are warmly intoxicating and you will not be able to stop reading, because what lies on the next page could bite you. It is also filled with vivid descriptions of parasites, which are some of the world's most intriguing creatures, in quite detailed scientific voice. Dr. Kwa weaves the life of these parasites with his own life journey from Malaysia to Australia to Malaysia and to the United States. This is definitely a book to be voraciously read by young and old, particularly those whose imaginations can and will run wild.

When I read the *Parasite Chronicles*, I caught my breath, because I was reminded of deep, meaningful conversations Dr. Kwa and I had over the years. I was also reminded of how fortunate I was to have taken Dr. Kwa's course and been his mentee as a graduate student, as these interactions launched me on a wonderful and joyous career filled with studying parasites. It has also allowed us to become colleagues and friends, a relationship that I treasure. Dr. Kwa and I both stepped out of Malaysia, making footsteps, each of us. I think it is important and poignant that he has made it so clear that we should not expect a return for challenges met and overcome but just pure gratitude and satisfaction for our footsteps taken. We are happy because of the dreams we follow. Dr. Kwa's joy in his professional and personal life is evident in the pages of this book. Through this lens he takes his reader on a remarkable journey using parasites as the vehicle.

Protein Potential LLC, Rockville, MD,   B. Kim Lee Sim
USA

Sanaria Inc., Rockville, MD, USA

Foreword by Stephen L. Hoffman

We have had vaccines to prevent diseases like smallpox, typhoid fever, tetanus, diphtheria, rabies, polio, measles, mumps, and German measles for 50–200 years. Viruses and bacteria cause these diseases. However, in 2017 we do not have a single licensed and marketed human vaccine for diseases caused by parasites. These are diseases with exotic names like elephantiasis, river blindness, Baghdad boil, kala-azar, and bilharzia and diseases that are so common that we all know of them—malaria, tapeworm, and hookworm. We do not have vaccines for parasites, because parasites are much more complex and larger than viruses and bacteria, making it much more difficult to use a vaccine to train the human immune system to attack and destroy them.

In the *Parasite Chronicles*, Professor B.H. Kwa takes the reader on his half-century odyssey with parasites and in the process leaves us with a much clearer sense of the identity and impact of these complicated, often weird organisms, and how we become infected with them. This is a journey that took him from the town of Taiping in the state of Perak in peninsular Malaysia to the University of Malaya in Kuala Lumpur to the Australian National University in Canberra back to Kuala Lumpur and then on to Harvard and finally and currently to the University of South Florida in Tampa. It is a journey that took him from parasitology to biomedical science and eventually to public health in the general population and ultimately to global health. It is a story of how a small-town boy from Malaysia ended up as the founding chair and professor of Global Health and associate dean at a university in Florida. In other words, it is the archetypical story of how a small-town boy makes good, the story of how we are only limited by our vision and work ethic, and how much we need our immediate family.

However, at the core of this narrative, he vividly and entertainingly explains to us about these often bizarre, sometimes romantic, but often deadly parasites, while intertwining his remarkable professional and personal journey. His riveting stories teach us of their impact on Napoleon, Julius Caesar, the ancient Chinese, and sadly the 500,000 children who died of malaria last year. As stated in the preface, "This book is aimed at a wide and eclectic audience." Having read the *Parasite Chronicles*,

I am sure that whoever reads this book, regardless of age or walk of life, will take with him or her a deep sense of and appreciation for these phenomenal creatures and for the man, Boo Hoe Kwa, who has weaved together a wonderful tale of himself, his family, and parasites.

P.S. I met Boo Hoe Kwa and his graduate student, Betty "Kim Lee" Sim, on the same day at a biomedical conference in Kuala Lumpur. I have a lot to thank him for, as Kim Lee and I have been married for 29 years.

Sanaria Inc., Rockville, MD, USA   Stephen L. Hoffman

Preface

This book is aimed at a wide and eclectic audience: a new student starting out in public health, a premed who is unsure why he or she wants to be a doctor, the freshman in college still undecided whether to major in biology, an undergraduate taking a gap year to do some adventure travel in Central America, and the new graduate contemplating joining the Peace Corps. The book will also be of interest to the well-read traveler, the curious armchair tourist who never misses Anthony Bourdain on TV, the adventurous foodie going to Seoul or Singapore, and the ecotourist soon to be off to the Galapagos, Borneo, or the Serengeti. It will also be a great stocking stuffer for the holidays to the daughter, grandson, niece, or nephew planning on traveling to Thailand or Costa Rica next summer.

The idea of this book came about when, over the years, many of my students suggested that I should write down the anecdotes and stories I tell in my class on parasitology at the USF College of Public Health and before that at the University of Malaya. I believe that my job in the classroom is to make the subject of parasites interesting enough that the students would want to read more about the subject themselves, as there was no point in just repeating what is already in the textbook. I try to find really interesting snippets about certain diseases, or unusual aspects of a particular parasite, or bizarre features in their life cycle that will stick in the student's memory even years after they have graduated. It has always been gratifying to me when some of my students email me to say that they can still remember an aspect about Chagas disease, or a story about filariasis or hookworms that I had mentioned in class, a decade after they had left school. So I finally decided that I will write this book in the final year before I retired, in honor of the generations of students that I have taught.

I have organized the sequence of the chapters both to reflect a natural taxonomy of the different groups of parasites important to human disease and to relate my own life's journey into the world of parasites, hence the title of the book. Chapters 1–3 are on the cestodes and tapeworm diseases. It is a convenient introduction to parasite diseases as it is also a *seguè* into how I first became interested in the study of a tapeworm disease when I was in graduate school in Australia. It also relates to reports of how cestodes were used in weight reduction diets in the US Midwest in 1921, how tapeworm larvae may have played a role in the life of an important historical figure such as Julius Caesar, and its impact on peoples as different as Turkana tribesmen in Kenya to sheep herders in northern Manchuria.

Chapters 4–5 discuss the trematodes such as the liver flukes as well as snail fever. It relates how liver flukes affect integrated farming practices in SE Asia, how snail fever involved the lives of ancient Egyptians, its role in the civil war in China, and perhaps even the death of Napoleon Bonaparte. The author also reveals why he avoids eating raw fish of any kind. Chapters 6–9 are on the nematodes and describe how the author started on his lifelong study of lymphatic filariasis after he returned to Malaysia where he grew up. These chapters also discuss events such as the court case in Canada which involved attempted murderous use of parasites to endanger the lives of college roommates, hookworms and their role in theories about how the indigenous Americans first arrived in North America, River Blindness in the Amazonian Basin, Guinea worm disease brought from India to Malaysia, and the author's search for parasitic worms that infect human brains, in cobra-infested oil palm plantations.

Chapters 10–12 cover protozoan parasites. They relate about the parasite that is most responsible for why we see so many people nowadays carrying around bottled water, Charles Darwin's encounter with the bugs that carry Chagas disease, the 1933 parasite outbreak during the Chicago World's Fair, how a parasite from cats may manipulate human behavior, how water skiing on a lake during the hot summer months may get you a brain eating parasite, and how far we have progressed in finding the Holy Grail in parasitology—the elimination of perhaps the greatest of all the parasitic diseases that affects humans: malaria.

Tampa, FL   Boo H. Kwa
July 2017

Acknowledgements

I wish to thank Donna Petersen, Dean of the College of Public Health, University of South Florida, for encouraging me to write this book and for allowing me to take a year's sabbatical to write it. I am very grateful for her continuous encouragement, support, and help throughout this time and for making the initial contact with Springer Publishing. I am also very grateful to Lars Koerner, my editor at Springer, for his help and cheerful encouragement at all stages in the preparation of this book. He graciously welcomed me at his office in Heidelberg in Spring 2017, and our discussions over a wide range of subjects led me to conclude that I was very fortunate to work with such an enlightened publishing house.

Many colleagues and scientists had read various parts of the drafts of the manuscript and their helpful comments, enthusiasm, encouragement, and critiques were invaluable. I wish to thank Ricardo Izurieta, Mak Joon Wah, Jackie Cattani, Tom Unnasch, Kim Lee Sim, Steve Hoffman, Mike Pentella, Yehia Hammad, and Donna Petersen who had all contributed by letting me know that I was on the right track in balancing scientific content with an accessible narrative to reach a general audience. I wish also to thank all my students over the years who have encouraged and pestered me to write down all the stories I tell in my lectures in parasitology. I am grateful to Mak Joon Wah and Ann Vickery DeBaldo for the use of images in the illustrations and to Springer for permission to use images from the *Encyclopedia of Parasitology* (Heinz Mehlhorn, Editor, 2016) illustrated in this book.

I am also very grateful to family members who had also kindly "market tested" the book's readability to a nontechnical audience, and although I must have severely tried their patience by sending them excerpts of the book, they have suffered in good grace. I wish to thank my wife Lucy Tze and daughters Shiamin and Shialing, my brother Andrew, sister May, uncle Martin, and aunt Patricia; and to my mother, who was mystified that someone would need to take a year to write a book but nevertheless was very patient and quietly encouraging.

Most of all I wish to thank my wife Lucy Tze for putting up with my very disruptive writing style of waking up to write between 5 o'clock and 8 o'clock every morning and then harassing her to read and critique every preliminary draft I had written when she woke up. Then to tolerate the editing and rewriting that took up the rest of the day for the past year. Her continuous encouragement, support, love, patience, and good cheer made it possible for me to complete the book within the deadline I had promised to myself.

Contents

About the Author

B. H. Kwa is Emeritus Professor at the College of Public Health, University of South Florida, Tampa, where he was the Founding Chair and Professor of Global Health and recipient of the Distinguished Teaching Award; he had also served as Associate Dean for International Programs and was previously Chair of Environmental and Occupational Health. His other academic appointments included being W.H.O. Visiting Scientist at the Harvard School of Public Health; Nuffield fellow at Imperial College, London; Fulbright fellow at USM, Penang, Malaysia; and Associate Professor at the University of Malaya, Kuala Lumpur. He had graduated from the Australian National University, Canberra, and the University of Malaya, Kuala Lumpur.

Chapter 1
Parasites and Tiger Snakes

The Tiger Snake that lay on the board before me was of course already dead. Not being fond of snakes, I even took the extra precaution of decapitating it after it was thoroughly chloroformed and pronounced dead by the herpetological lab technician. I had proceeded to dissect its vividly colored skin and had pinned the skin down on both sides of the carcass with two neat rows of pins. Just then Mike came in cautiously to look over my shoulder to examine the glistening muscles of the freshly dissected carcass of the Australian Tiger Snake, *Notechis scutatus*, reputedly one of the world's most poisonous snakes.

"Are you sure that it is dead?" asked Mike skeptically. Now you must understand that Mike was born and bred in New Zealand, an island nation like Ireland, which by some evolutionary quirk, is free of any native snakes. So not only was he not fond of snakes like me, but he had not been near any snakes in the wild in his entire life. "I'm sure." I said. "Look, I think that long white thread-like strip right there may be a sparganum. Should I proceed?" As I cut into the snake's muscles around the parasite, the powerful fibers of the longitudinal muscles of the snake contracted violently, ripping the pins off like a giant zipper, causing the headless carcass to suddenly pull itself up from the board in a curving arc. Mike was at that very moment bent over my shoulder to have a better look, and the snake's carcass, no longer held down to the board, had risen up and curved towards my neck like a cobra's strike. I must have jumped up and let out such a scream that it was heard around the building. It became part of the folklore of the Zoology Department of the Australian National University (ANU).

This book describes a journey that started with my interest in biology and the ecology of living organisms as an undergraduate student, which later broadened into studying parasitism and parasites. I was fascinated by the myriad variety of intricate life cycles that parasites have evolved to ensure that they can thrive within the body and organ systems of other animals and to survive the powerful host immune responses mounted against them. Complex survival strategies in their evolution have allowed parasites to infect and colonize every known family of insects, fish, amphibians, reptiles and mammals. Parasites also have to evolve adaptive strategies

© Springer International Publishing AG, part of Springer Nature 2017
B. H. Kwa, *The Parasite Chronicles*, https://doi.org/10.1007/978-3-319-74923-5_1

to enter the host, exit the host and survive for extended periods outside the host as well. In most instances they also colonize secondary hosts such as snails, insects, invertebrates and smaller mammals, to allow them to spread to new hosts.

My personal journey continued from parasitology to biomedical science and eventually to public health in the general population and ultimately to global health, i.e. the area of public health writ large which deals with global populations not restricted only to the industrialized wealthy countries in North America and Europe. Many parasitic diseases such as malaria, schistosomiasis, lymphatic filariasis, onchocerciasis, and intestinal helminths occur mainly in the poorer parts of the world. These have become the neglected "orphan" diseases, as funding sources that support research have generally focused on the more pharmaceutically lucrative diseases of the affluent such as heart disease, cancer and diabetes which mainly result from a sedentary lifestyle and over-nutrition. Much of the meanderings and diversions along my journey were necessitated by personal, family, economic and career choices, not any different than most people. I do not claim any particular altruistic reason for choosing this path—but it had been a lot of fun!

I guess my interest in parasites all started when I was a graduate student at the Australian National University in Canberra, Australia. I had just started on my Master's degree in the department headed by the eminent parasitologist Prof. J. D. Smyth, and "Prof", as he was universally known by all in the department, had offered me a spot to work on a project on *sparganosis* when I had applied to work on one of the research projects under his research team. I jumped at the idea because the topic intrigued me after I had read the papers and book chapters on the reading list he had assigned, all relating to infections by the intermediate stage of tapeworms belonging to the genus *Spirometra*. There was this particular picture of a patient from Vietnam with a grotesquely inflamed eye that gave the appearance that the entire eyeball was dropping out of his eye socket (Faust and Russell 1964, Fig. 235, p 646). The eye was infected by the sparganum parasite. It was an unforgettable image. "Prof" was a brilliant amateur motivational psychologist, as he knew that a 22 year old graduate student from Malaysia would be instantly captivated by such a bizarre disease that I had never even heard of, although it was from a country in Southeast Asia not more than a few hundred miles north of where I grew up.

A tapeworm belonging to the genus *Spirometra* first starts its life as an egg with a tiny domed lid, called an operculum. When the egg hatches in water, the operculum pops open and releases a microscopic ciliated larva, called a coracidium, which swims around until it gets eaten by a tiny crustacean called a copepod. It then grows into a larva called a procercoid within the copepod. Now it gets interesting. If the copepod in turn gets eaten by say a fish, frog or toad, or even a tadpole, the procercoid then grows into a thread-like larva called a plerocercoid larva (Fig. 1.1). When the tadpole grows into a frog, the plerocercoid larva will simply continue to hitch a ride in the muscles of the grown frog, otherwise not causing it any apparent harm. If the frog, for instance, is then eaten by a snake, the plerocercoid larva will grow in length but will not develop any further in the snake. This feature of the plerocercoid larva's ability to be shuttled from intermediate host to intermediate host is called "paratenic" and contributes to its survival. Only when the frog or snake

Fig. 1.1 Sparganosis (Source: Encyclopedia of Parasitology Ed. Melhorn 2016, Fig. 1)

hosts are ultimately eaten by a mammalian carnivore, say a fox, dog, or civet cat, will the plerocercoid larva proceed to develop and mature into an adult tapeworm in the small intestines. The mechanism by which the plerocercoid larva is stimulated to develop into a mature tapeworm is found in the specific combination of different types of bile salts secreted into the small intestines of mammalian carnivores which are different from those in the intestines of the other intermediate hosts. This is one example of how a parasite uses such physiological triggers of its host to initiate developmental stages in their life cycles as they move from one host to another.

When this plerocercoid larva was first described by the famous biologist and parasitologist Patrick Manson in 1882 in Amoy, China, it was the common practice to call this larval stage a sparganum. Human infections can occur when infected copepods are accidentally ingested in drinking water or if the plerocercoid larva (sparganum) is eaten in raw fish. The larva will penetrate the gut and work its way through muscle and connective tissue, instead of developing into an adult tapeworm, because human bile salts resemble more closely those of the other intermediate hosts. The person then becomes one more "paratenic" host for the sparganum. Most human infections result in minor clinical conditions or may remain asymptomatic. Sometimes the sparganum may accidentally migrate into the eye of the unfortunate infected person and cause severe ocular complications. But that is relatively rare.

Instead, the Vietnamese patient, whose image I first saw in the book, had contracted the eye infection in an even more bizarre and engrossing fashion. Among certain communities in rural Indochina, the raw flesh of frogs were used as poultice for eye trauma, the way raw meat is used for a black eye, say, in Chicago after a bar brawl. If the flesh of the frog, used in this way, happened to be infected by a sparganum, the parasite would be stimulated by the body temperature of the person and would start migrating towards the source of warmth, in this case the eye. The sparganum may migrate under the conjunctiva and form a nodule or may even migrate into a retrobulbar site, meaning behind the eyeball, and proceed to cause inflammation and pressure that push the eyeball out of its socket. The appearance of the wretched Vietnamese villager, in the picture that had mesmerized me, had a retrobulbar infection.

The question that "Prof" Smyth wanted me to answer was how the sparganum of *Spirometra erinacei*, which is the species found in Australia, was able to penetrate the wall of the intestines of the host so rapidly, as well as how it could migrate apparently through connective tissue and muscles so effortlessly. "Like a hot knife through butter" as he put it in his soft Irish brogue. The sparganum is soft and flaccid, resembling wet cotton thread, and does not have hard appendages like some of its cousins (oncospheres and the emerging larva of some other families of tapeworms are armed with tiny hooklets, resembling grappling hooks, that tear their way through the soft tissue of the host). Furthermore, microscopic examination does not even show the presence of visible glands that produce histolytic (tissue-dissolving) enzymes that dissolve tissue like those of another group of parasites, the hookworms.

As "Prof" was already mentoring several other graduate students and had his hands full, he assigned me to be mentored by Mike Howell, then a young earnest veterinarian from New Zealand with a PhD in parasitology, who had recently started as Lecturer in Zoology at ANU and was looking for a student to work on a research project under his supervision. It was fortuitous for my research question that Mike had worked on liver flukes previously, a group of parasites distantly related to the tapeworms, but unlike most tapeworms do not have larval stages with hard physical appendages such as hooklets which have been evolved to mechanically tear through host tissues. So it was a scientific question that he must also have thought about: are there other ways that parasite larvae have evolved to migrate in host tissues?

To begin my project, Mike put me to work dissecting scores of Cane Toads, as the toads had been reported to be a common intermediate host of the sparganum in Australia's northern sub-tropical state of Queensland. To start on my research, I had to first harvest enough of the parasites so that I could have a large enough sample of live specimens available to use in laboratory experiments. I was very fortunate to start on my career in parasitology under the mentorship of Mike Howell who was invariably friendly, encouraging and patient with me as his first graduate student.

The story of how the Cane Toad *Bufo marinus* (since renamed *Rhinella marina*) was introduced into Australia was itself a salutary lesson of how attempts at "biological control" can sometimes have catastrophic consequences. Cane Toads were deliberately introduced to Australia from Central and South America to control the cane beetle *Dermolepida albohirtum,* which was a pest threatening the sugar cane industry in Queensland. However because the toad did not have any natural predators in Australia to control its population, they soon proceeded to proliferate spectacularly throughout coastal Queensland, the Northern Territory and northern New South Wales, covering a huge swath of the entire northeastern portion of the continental landmass of Australia. Female toads can lay up to 25,000 eggs at once and the strings of eggs have been reported to stretch up to 60 feet in length. As a defense mechanism, these toads have glands that secrete a milky white toxin which is extremely poisonous, and the spread of these toads constitutes an ecological disaster as several native species, such as native goannas or monitor lizards (*Varanus panoptes*) and indigenous quolls or marsupial cats (*Dasyurus hallucatus*), face extinction due to lethal toxic ingestion of the toads. Dogs have been known to die just licking the skin of the toads. The irony is that the toads themselves eventually proved to be ineffective in controlling the cane beetles, for which they had been introduced, as most of the beetles live among the cane leaves too far above where the toads can reach them.

Soon after dissecting the first thirty or so toads, I could only recover two live spargana (plural of sparganum) for my experiments. It became obvious that at that rate, I would not likely complete my thesis research in the allotted 2 years expected of me. I would never have recovered enough spargana and the consignments of live Cane Toads from Queensland would quickly eat up all the funding on my small research budget. Mike then came up with the brilliant idea for me to look for spargana in snakes instead. The thinking was this: if frogs, toads and snakes were all "paratenic" hosts, then we should examine the predator at the top of the food chain, in this case the snake. Since the average fully grown snake would likely have fed on scores of frogs and other amphibia in its lifetime, there must be a lot of spargana accumulated in its body just waiting to complete their life cycle in the final host, the mammalian predator.

When Mike made this incisive intuitive leap in deductive reasoning, he must have instantly regretted introducing the dreaded snake into the research project of his novice graduate student. He was now responsible for supervising an inexperienced graduate student, one who surely could not be completely trusted to handle poisonous snakes. After all, Australia was reputed to have the most number of native species of poisonous snakes of any continent and many of the live snakes that were brought into the lab that autumn were likely to be poisonous. For a Kiwi like Mike, this must be a nightmare.

Mike was right in his hunch however, as the very first snake I dissected had close to thirty live spargana living within its muscles. Thus, with a steady supply of spargana from the snakes, my experiments were off to a good start. Most of the snakes were caught in the countryside across the border from Canberra in northeastern New South Wales, and they harbored enough spargana for the duration of my project. I did not cause any further mishaps, but noticed however that since my first clumsy encounter with that snake, Mike had thought better and had stopped coming to the lab whenever I had a fresh consignment of snakes to dissect!

We discovered that the sparganum did indeed have cells producing powerful tissue-dissolving enzymes, but instead of visible glands in the scolex (head) of the parasite secreting them, it appeared that the sparganum's surface, or tegument, have cells that produce the proteases to cover its entire body. The cells were only apparent when examined under an electron microscope, which magnifies objects too small for the light microscope to see (Fig. 1.2). Electron-microscopic examination showed that the entire tegument is studded with these enzyme producing cells. Thus the entire body of the sparganum is able to dissolve the surrounding host tissue that it

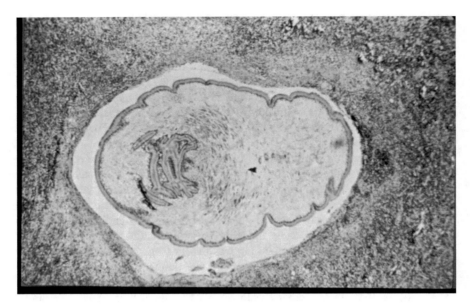

Fig. 1.2 Sparganum, histological section in mouse tissue capsule. Picture by author

migrates through, exactly "like a hot knife through butter" and it explained how it can travel via the food chain from one intermediate host to another so efficiently.

***

It was also during this period of my life that I met my wife Lucy Tze who was then an undergraduate living in Bruce Hall across the beautiful campus green of ANU, studded with fragrant Australian eucalyptus trees, from Burton Hall where I lived during my undergraduate and graduate years in Canberra. She is my constant companion, guide and friend throughout my travels and misadventures as related in this book.

***

The sparganum belongs to the genus *Spirometra*, of which there are several species such as *Spirometra mansoni*, *S. erinacei*, *S. mansonoides* etc. Sometimes parasites interact with their hosts in unexpected ways. Although the term "parasite" suggests that the host must inevitably suffer a nutritional deficit since it must support another living organism within its body, rats infected with *S. mansonoides* for example will actually gain weight, be larger and heavier but are otherwise normal in behavior. *S. mansonoides* secretes a factor similar to the growth hormone of rats and it had been speculated that perhaps this is an example of how parasites can sometimes manipulate their hosts to their own selective advantage for survival of their species. It has been suggested that larger and heavier rats are less agile, and will be more easily caught and eaten by cats or similar predators in the wild, helping to ensure that the spargana of *S. mansonoides* will reach their final host and develop into adult tapeworms and complete their life cycle. There is now a growing body of

Fig. 1.3 Diphyllobothrium latum (Source: Encyclopedia of Parasitology Ed. Melhorn 2016, Fig. 2)

evidence that such manipulation of hosts by parasites is not uncommon among parasite species, and we will see many other examples of this later.

The *Spirometra* parasites belong to a taxonomic order of cestodes (tapeworms) called Pseudophyllidea. These parasites require an aquatic environment, as part of their life cycle is completed in water and various species have adapted to both freshwater as well as marine environments. One such species, *Diphyllobothrium latum*, was responsible for a spate of tapeworm infections among elderly Jewish ladies in New York City in the 1930s (Fig. 1.3). This fascinating tale was humorously described by Robert Desowitz in his book "New Guinea Tapeworms and Jewish Grandmothers". It relates how immigrant fisherman in the lakes of the American Mid-West had brought their occupation, as well as their tapeworms, from Scandinavia and introduced *D. latum* infections to fish in the lakes around Wisconsin and Minnesota. However it was the advent of refrigerated trucks and railroad cars that enabled the fresh fish trade from the Mid-West to reach cities like New York half a continent away. Then it was the art of making gefilte fish, usually by the matriarch periodically tasting to test for "doneness" of the fish that ensured that she was the only one in the household who ingested undercooked fish with the infectious plerocercoid! Such elegant detective work enlivens the study of parasite transmissions.

Treatment of human diphyllobothriasis is with praziquatel and niclosamide. With human sparganosis, surgical removal is indicated although long term treatment with mebendazole or praziquantel had mixed results.

***

Another related tapeworm *Schistocephalus solidus* has been shown to be one of the classic examples of a parasite which manipulates its host to enhance its survival advantage (Fig. 1.4). The maturation of the plerocercoid larval stage is dependent on temperature and "Prof" Smyth, who was a pioneer in the in vitro culture of cestodes as well as one of the first to conduct research on *S. solidus*, had demonstrated that

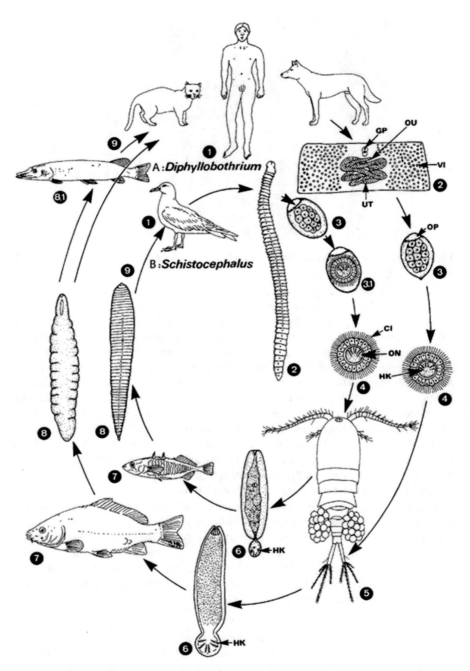

Fig. 1.4 Diphyllobothrium latum (Source: Encyclopedia of Parasitology Ed. Melhorn 2016, Fig. 1)

once it approaches 35° C the *S. solidus* plerocercoid will start to mature and grow spectacularly in size, sometimes growing to 90% of the weight of the fish intermediate host. In Europe the stickleback *Gasterosteus aculeatus* is a common host for the parasite and in summer, as water temperatures rise, the parasite will grow rapidly in size and consume an increasingly larger proportion of the fish's energy resources. This in turn appears to drive the fish to seek more food in warmer waters, become more solitary and adopting riskier behavior which exposes it to predation by birds. Fish carrying the large parasites become distended and have their swimming gait altered, becoming slow and more erratic swimmers and therefore more likely to fall prey to the birds. Thus all these factors increase the survival of the parasite by increasing their likelihood of reaching the final bird host which is required for complete parasite maturation.

Chapter 2
Tapeworms Down Under and Elsewhere

While I was a graduate student in Canberra being introduced to the world of parasites, Australia was then battling more important parasitic diseases than the sparganum which I had studied, which affected only small communities in SE Asia. The sheep industry was very important to Australia's economy and it faced several serious tapeworm diseases at that time. One of those was hydatid disease (or cystic echinococcosis) which is caused by the tapeworm *Echinococcus granulosus* (Fig. 2.1). This was a serious agricultural, medical and public health problem in Australia when I was a graduate student in the 1960s and it still remains the sort of public health issue that causes local health and veterinary authorities in Australia to post periodic health advisories in rural and farming communities. As tapeworms go, the adult tapeworm of *E. granulosus* is relatively small, only 6–7 mm (around a quarter of an inch) long, and lives in the small intestines of domestic dogs and in wild canids like wolves, jackals, hyenas and coyotes in different parts of the world. The scolex of the tapeworm is embedded firmly among the villi in the small intestines of the host by four suckers and a crown (or rostellum) of 28–50 hooks. There the adult tapeworm will mature and the segment (or proglottid) at the posterior end will produce and release eggs throughout its life in the intestines. The eggs are passed out with the dog feces and if they are swallowed by sheep, or the unfortunate human incidental host, a tiny oncosphere or hexacanth larva (so called because it has three pairs of tiny hooks) will hatch out and penetrate the gut of the sheep using the hooklets like the *pitons* of rock climbers to migrate in the tissues. They will eventually enter a blood or lymph vessel and be carried in the blood stream or lymphatic system, usually to the liver where the young embryo will proceed to grow within a cyst known as a hydatid cyst (Fig. 2.2). For the cycle to complete, the cysts will have to eventually be ingested by the canine host, where the hydatid cyst will evaginate (unfold), the scolex attach to the mucosal surface of the intestine of the dog and grow into a new tapeworm. Notice here that the Cyclophyllidean cestodes described here bypass the need for an aquatic stage and had evolved a life cycle that allows them to pass directly from intermediate to final hosts.

© Springer International Publishing AG, part of Springer Nature 2017

B. H. Kwa, *The Parasite Chronicles*, https://doi.org/10.1007/978-3-319-74923-5_2

Fig. 2.1 Echinococcus
species (Source:
Encyclopedia of
Parasitology Ed. Melhorn
2016, Fig. 2)

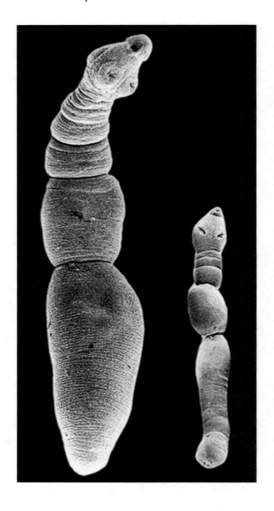

In the sheep intermediate host the hydatid cyst will grow to about 1 cm in size, and is essentially a tapeworm scolex protected by a fluid-filled sac surrounding it. Along the inner surface of the wall in the hydatid cyst, new protoscoleces will grow by a budding process, and as some of them grow larger they in turn may become new "daughter cysts" within the primary cyst. Rupture of the cyst can be life threatening, as the sudden release of parasite proteins in hydatid fluid may cause anaphylactic shock. Moreover, the daughter cysts that spill out will each develop into new secondary hydatid cysts wherever they become implanted in the body. There have been reports that during surgery to remove the cyst, accidental ruptures had seeded the entire abdominal cavity with new daughter cysts. This is therefore a particularly nasty and dangerous parasite (Fig. 2.3).

Human infections occur most commonly in the liver, but may also be found in the lungs, muscle, bones, kidney, spleen and brain in that order of likelihood of

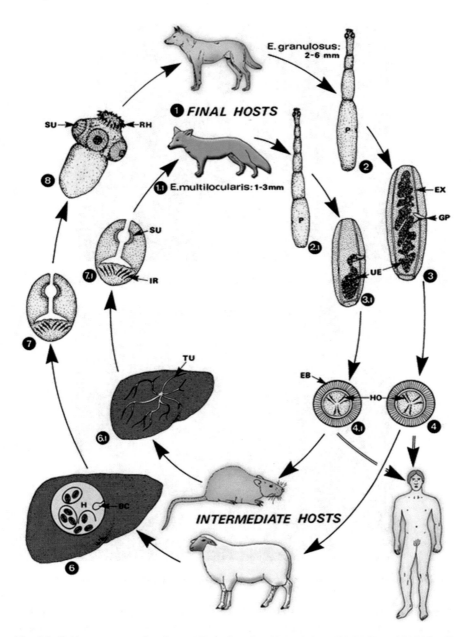

Fig. 2.2 Echinococcus species (Source: Encyclopedia of Parasitology Ed. Melhorn 2016, Fig. 1)

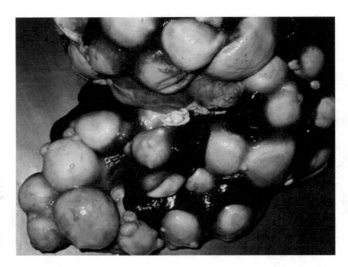

Fig. 2.3 Echinococcus species (Source: Encyclopedia of Parasitology Ed. Melhorn 2016, Fig. 7)

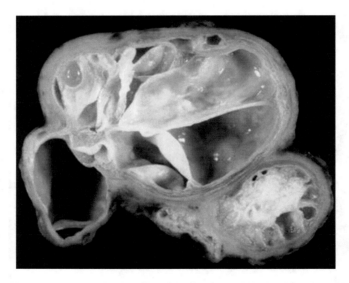

Fig. 2.4 Echinococcus species (Source: Encyclopedia of Parasitology Ed. Melhorn 2016, Fig. 6)

occurrence. The symptoms result from physical pressure exerted by the growing cyst and therefore will vary with the number and size of the cysts, as well as the organ in which they are located (Fig. 2.4). The vast majority (around 80%) of *E. granulosis* infection results in single cyst infections, and the growing cyst will mimic the pathology of a benign tumor. However the results can be severe depending on where they occur. Liver infections can cause hepatomegaly (abnormal enlargement of the liver) or obstructive jaundice; lung infections can cause hemoptysis (coughing up blood) and difficulty in breathing; if the cyst grows in the vertebrae it can cause paralysis of the limbs (paraplegia); and kidney infection can cause various forms of renal disease. A growing cyst in the brain, though more rare, can result in very serious neurological complications including blindness. Another type of hydatid cyst is known as an osseous hydatid. This can occur if the embryos develop in the bones where they grow into extended protoplasmic strands within the cancellous tissue, which is the soft spongy tissue in long bones, where they weaken and damage the bone tissue including those of the pelvic arch and vertebrae, in addition to those of the long bones.

Wherever dogs and sheep live together in close proximity, it is easy to see that dog feces will contaminate the pastures where sheep graze. In countries where sheepdogs are an important part of the sheep rearing industry, such as Australia, New Zealand, several countries in southern South America, North Africa, the Middle East, South Africa and Central Asia and North West China, an endemic sheep-dog cycle can become well established. Hydatid disease is an important public health issue because human infections are unfortunately still common in rural areas where sheep are raised. Sheep herders are contaminated by the infectious eggs which frequently attached to the dog fur and children playing in the contaminated pasture would pick up the eggs in their hands and accidentally ingest them that way. It is important to remember that humans are only infected by the tapeworm eggs shed from dogs, which is the only source of infection and human infection is not related to eating lamb, mutton or sheep products.

In poorer less developed communities, sheep are still commonly slaughtered in the farm and the discarded offal (organ and gut tissue) are fed to the sheep dogs, While this may be an economical way to reduce the expense of buying feed for the working dogs, it ensures that the intermediate cyst stage in the sheep re-enters the dog, to grow into the adult tapeworm and thereby maintaining the life cycle of the parasite. While this practice is still not uncommon in poorer countries today, it is no longer a source of infection in the modern farm in developed countries. The advent of effective pharmaceutical treatment of the sheepdogs with drugs such as praziquantel, have largely contributed to the elimination of hydatid disease in sheep farming. Other control measures include good animal husbandry such as the proper disposal of sheep viscera and carcasses of dead sheep, and the removal of carcasses of wildlife that may harbor the parasite, to prevent access to dogs eating them. Simple public health practices such as hand washing after handling dogs have also played an import role in eliminating hydatid disease.

Among the nomadic Turkana community in the arid water-scarce desert of northwestern Kenya a uniquely intimate relationship with their dogs occur, whereby the dogs perform duties including licking and cleaning the cookware and food serving utensils, cleaning up the excreta and vomit of the pre-toilet trained children, and consuming the menses of the women. Dog feces were also used medicinally and cosmetically. The dogs were routinely fed the carcasses and entrails of the slaughtered livestock and as much as 65% the canines are infected. Therefore exposure to echinococcus eggs is very high among the Turkana people and

Fig. 2.5 Echinococcus
species (Source:
Encyclopedia of
Parasitology Ed. Melhorn
2016, Fig. 10)

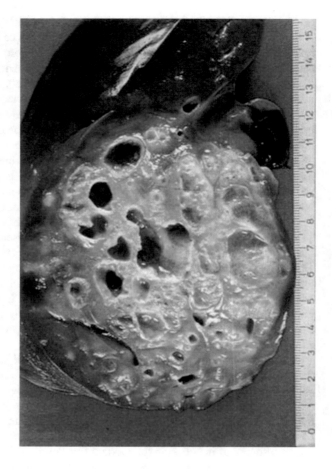

prevalence the rate of hydatid disease is between 7% and 10%, among the highest ever recorded.

Echinococcosis may be caused by several other species of Echinococcus, although some of these are very rare. Human infections may be commonly divided into two forms, cystic echinococcosis (CE) and alveolar echinococcosis (AE), because besides *E. granulosus* which causes CE there is an even more dangerous form caused by a related parasite called *Echinococcus multilocularis*. The larval stage of *E. multilocularis* which causes AE does not mature into cysts as the limiting membrane is not formed, so that the unbound larval vesicles will proliferate in different directions and invade surrounding tissues. While the liver is the most common organ infected, the vesicles may also metastasize via the blood stream to other organs such as the lungs or brain, acting essentially like a malignant tumor, making AE extremely dangerous (Fig. 2.5). Mortality is between 50% and 75% in infected persons, because it mainly occurs in underserved indigenous communities living in remote northern latitudes of Asia, Europe and North America (think of

Siberia, Lapland and the Yukon) where foxes, wolves, coyotes and dogs are infected with the adult tapeworms and rodents are the natural intermediate hosts. Infections in humans are commonly through food or water contaminated with the parasite eggs most likely from infected working dogs such as sled dogs.

As someone who grew up in the tropics in Malaysia, it was difficult for me to imagine the world of sled dogs, blizzards and frozen tundra, harsh conditions that promoted the propagation of AE. Among indigenous communities living in conditions of extreme frigid night time temperatures, working dogs in winter are commonly sheltered indoors with the human inhabitants at the end of the day, and in isolated farms without electricity people often eat dried preserved meats and smoked fish without cooking due to the paucity of scarce fuel. Such close proximity between dogs and humans ensures likely human ingestion of *E. multilocularis* eggs from dog fur contaminating the food which is eaten uncooked.

I had my opportunity finally to visit such a place during winter to experience for myself the conditions under which AE will most likely be transmitted. In January 2013, I was invited to deliver a paper at a conference in Harbin, the capital of Heilongjiang Province in northeast China, to commemorate the contributions of Dr. Wu Lien-teh (1879–1960) who treated patients and instituted public health measures to eradicate a devastating epidemic of the bubonic plague in what was then Manchuria in 1910–1911. Wu was a physician and biomedical scientist who was born in Penang, Malaysia, where he was educated at the prestigious Penang Free School, went on to England on a scholarship where he graduated in medicine from Cambridge University and later trained under Sir Ronald Ross at the Liverpool School of Tropical Medicine and Elie Metchnikoff at the Pasteur Institute in Paris. He was later nominated for the Nobel Prize in Medicine for his innovations in epidemic control and public health measures. Although he did not win the Prize, he was the first and only Malaysian to be nominated for that honor, and it is a source of pride to many Malaysians of my generation.

When I landed in Harbin that day, the temperature was -36° C and the Songhua River that flows through the city was frozen solid, pushing large riverine cruise ships and cargo vessels several meters into the air at dizzying angles. Harbin is in the same latitude as Siberia and the frozen tundra outside the city looked like a scene from "Dr. Zhivago" (Fig. 2.6). A few hundred kilometers to the northwest of the city near the Russian border is a region where there have been numerous recent cases of both CE and AE reported, and in fact two of the major hospitals in Harbin are prominent centers for the diagnosis and treatment of echinococcosis. My visit to the hospital affiliated with Harbin University was founded by Wu Lien-teh and is today well known for research in echinococcosis. I feel humbled to have had the opportunity to honor the important contributions he made to public health. The fact that he grew up in Penang, where I completed my high school education as well, was a source of pride and inspiration. For a parasitologist, it was also a wonderful opportunity to experience firsthand the harsh winter conditions under which AE is most likely to be transmitted in that part of the world.

Two other Echinococcus species that causes human disease, in a form known as polycystic hydatid disease, occur in Central and South America and they are

Fig. 2.6 Author on the frozen Songhua River, Harbin, 2013

E. vogeli, which has a wild canid-rodent cycle and causes the formation of vesicles in various human organs, mainly the liver and associated visceral organs such as lungs, stomach, heart, diaphragm, omentum, mesenteries and intercostal muscles; and *E. oligarthrus* which has a wild felid-rodent cycle with the infections in similar locations in the human viscera. These two species are however extremely rare, and occasional cases have been reported from Panama, Colombia and Ecuador. However, because the cases were mainly from rural or remote forest communities, where medical facilities are rudimentary, they could be under reported or misdiagnosed, making it difficult to have reliable estimates of their true incidence.

***

During the time I was a student in Australia, in the 1960s, I remembered that public health education regarding hydatid disease very much emphasized the need to prevent sheep farmers from feeding sheep viscera to dogs, or at least to boil the offal to kill the cysts first. That was a very sensible public health measure as many of the anti-helminthic pharmaceuticals such as albendazole and mebendazole for treatment of human echinococcosis prescribed today were not yet widely available at that time.

However during the period as a graduate student, I quickly learned that changing the behavior of farmers who had been practicing their way of life for generations was extremely difficult, especially when they felt that they were being lectured to by condescending health authorities. (Later I was to learn once again, when I returned home to Malaysia and started my research on lymphatic filariasis, that crafting a socially and culturally sensitive educational program is essential for any public health program to succeed). Aussie farmers also have a well-deserved reputation of being rugged individuals who are proud of their ability to take care of themselves. However the economics of running a sheep farm profitably in those days included not discarding farm waste which could be a food source for their many working dogs. So it turned out that one of the most effective and successful public health educational films produced during that period for hydatid disease prevention campaigns, did not only emphasize protective and safety issues but also appealed to the altruism of the farmer. The narrative used was that by boiling sheep offal thoroughly before feeding them to the dogs, it would prevent *innocent children from adjacent farms* becoming accidentally infected. When pilot educational reels were shown to test audiences, this was shown to be many times more effective in modifying the behavior of the farmers, compared to an alternative film which emphasized that boiling the offal would protect the *farmers themselves* from being exposed to hydatid disease. I was then a zoology student and had not yet realized that I was already being introduced to one of the proven tenets of public health education and health promotion. The idea of promoting altruistic behavior regarding exposing second hand smoke to family members was used with great effect in the tobacco cessation campaigns later in the United States in the 1970–1980s. It was to be another decade later before I was introduced into the whole world of global health and was to embark on a fulfilling and stimulating career in a school of public health in the United States.

As a personal *mea culpa*, I confess that I used this bit of self-administered psychology of altruistic behavior to quit smoking many years ago. I had started fooling around with cigarettes in high school and by the time I was a freshman in college I was a pack-a-day nicotine addict. I tried quitting at least five or six times but like other smokers before me, I was back to my bad habit within days. On one occasion I actually held out for more than a month but ultimately succumbed back into the habit and like others I made excuses about stress, peer pressure, and social norms to explain my recidivism. Finally, after I had returned to Malaysia, on the evening while my wife was in labor with my first child, as I paced outside the maternity ward of University Hospital in Petaling Jaya, I stubbed out the final cigarette, crumpled the rest of the unsmoked pack of cigarettes and threw it into the drain. With that symbolic gesture I silently pledged never to expose my first born to second hand smoke ever. It actually worked! After that whenever I started to weaken, and nicotine addiction is one of the most powerful of all addictions, all I had to do was to look at my daughter or to picture her innocent baby face in my mind, and that would give me enough strength to overcome the urge to smoke. To pick up a cigarette again would have felt like a major betrayal somehow, and I never touched another cigarette since that November night in 1973.

Chapter 3
Cysticercosis: The Ides of March

Unlike the tiny adult tapeworms of Echinococcus, tapeworms that belong to the genus *Taenia* are the archetype of what we think of as tapeworms. Human infections are associated with three species, *Taenia solium*, *T. saginata* and the more recently discovered *T. asiatica*. These are huge parasites, for which the name tapeworm was first used to describe them. The adult stage of Taeniid tapeworms live in the human small intestines with their scolex embedded in the mucosal wall of the intestine (Fig. 3.1). The main body of the tapeworm, known as the strobila, is flat and broad like a measuring tape and it can extend to 25 m (80 feet) long, and lives floating in the nutrient rich broth of the host intestinal lumen where it shares its food. This type of infection by the adult tapeworm is called taeniasis, and it seldom causes anything more life threatening than bowel irritation and discomfort. However, since sections of the strobila are occasionally passed out in the stool, still visibly moving and undulating, this must be disturbing to the carrier to say the least, and these large parasites must have been easily observed and known by our early ancestors (Fig. 3.2).

Thus the lineage of this tapeworm goes back at least to the ancient Egyptians. Mention was made in the Ebers papyrus (1500 BCE) where descriptions of the adult tapeworm had been recorded. Taenia eggs and calcified cysts of the larval stage were also discovered in the intestines and viscera of mummies, and early Egyptian physicians have written that they treated tapeworm infections with pumpkin seeds (*Cucurbita pepo*). Jewish religious dietary laws (600–500 BCE) as well as the Christian Old Testament had forbidden the consumption of swine for being "unclean" (Leviticus 11: 7.8) and later Islamic laws promulgated by Muhammad (570–632 CE) similarly prohibited the eating of pork and handling of swine products, perhaps related to knowledge of the prevalence of *Taenia solium* in pigs in the Eastern Mediterranean region.

It is intriguing to speculate that knowledge, or at least the suspicion, of a connection between pigs and cysticercus infection, caused by the larval stage of *Taenia solium*, must have occurred at least a couple of millennia ago. Aristophanes (448–385 BCE) had a character mention in one of his plays ("The Knights") that the

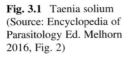
Fig. 3.1 Taenia solium
(Source: Encyclopedia of
Parasitology Ed. Melhorn
2016, Fig. 2)

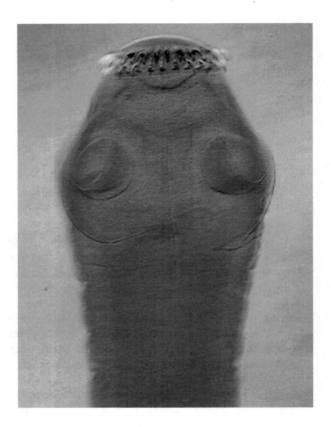

tongue of a person could be examined to see if it is "measled" like that of the pig.
Aristotle (384–322 BCE) actually described the presence of bladders or cysts (live
cycticerci) in pig muscle and compared some to hailstones (dead calcified cysticerci). He also observed that free roaming pigs had the cysts but not young nursing
pigs, making him one of the earliest scientists to have enquired into the natural
history of parasitic disease transmission. These referred to the presence of the
cysticercus larvae of *T. solium*.

Taeniid tapeworms belong to the taxonomic order Cyclophyllidea and all three
species which infect humans, *Taenia saginata*, *T. solium* and *T. asiatica* live as an
adult tapeworm in the human small intestines where this stage is nowadays easily
treated with pharmaceutical agents like praziquantel or niclosamide. While the
scolex of *T. saginata* attaches itself with four prominent round suckers (Fig. 3.3),
those of *T. solium* and *T. asiatica* have an additional crown of two circular rows of
tiny hooks in front of the four suckers, ensuring an even more secure anchor for the
main body of the strobila. The strobila itself is divided into as many as a thousand
segments known as proglottids which contain both male and female reproductive
organs. Tapeworms are therefore hermaphrodites and can live solitarily and reproduce. Since the male and female reproductive organs of the proglottids mature at

Fig. 3.2 Taenia solium
(Source: Encyclopedia of
Parasitology Ed. Melhorn
2016, Fig. 3)

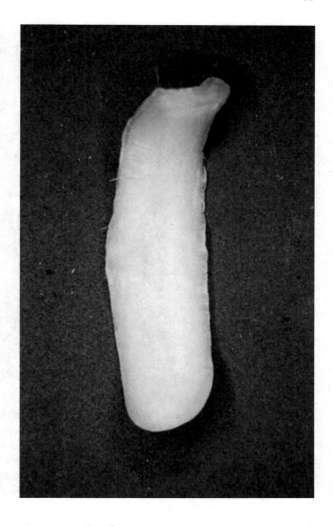

different times, the males first, they can reproduce by fertilization of ova from mature ovaries of a proglottid further down the strobila by spermatozoa from a different proglottid nearer the scolex. This is achieved because the strobila is folded in the host intestines, with mature male and female proglottids juxtaposed in close contact against each other, even though they may be actually several meters apart along the length of the strobila. Fertilized eggs pack the uterus of the mature proglottids which become essentially little packages of fertilized eggs ready to leave the host intestine and infect the intermediate host. These so called gravid proglottids will periodically break off from the ends of the strobila and pass out in the stool, occasionally crawling directly out of the rectum. These pieces of tapeworm can move by contracting and expanding, mechanically crawling like leeches on any surface they happen to be on. When I was small I heard horror stories of how

Fig. 3.3 Taenia saginata
(Source: Encyclopedia of
Parasitology Ed. Melhorn
2016, Fig. 1)

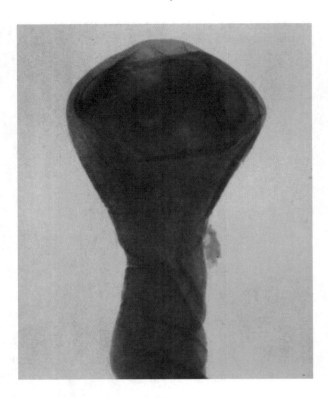

tapeworms have been found crawling on the bed of naughty boys who spent their pocket money buying unhygienic street food instead of coming home for supper!

For the tapeworm to continue its life cycle, the eggs must somehow reach its intermediate host. In low socioeconomic communities around the world, defecation in the open environment is commonplace. Under such conditions human tapeworms will thrive (Fig. 3.4). The gravid proglottids passed in the stool will eventually disintegrate and liberate the eggs into the soil. Grazing pigs and cattle ingesting the eggs from contaminated garbage or pasture will be infected when the eggs hatch in their duodenum after being partially digested in the stomach, liberating the oncosphere (or hexacanth) larva with its 3 pairs of tiny grappling hooks. The oncospheres migrate through the tissues of cattle to reach the muscles and in the case of pigs, numerous other visceral organs in addition to the muscles, and there develop into the cysticercus larva (Fig. 3.5).

Whereas the adult tapeworm of all three human Taenia tapeworms only cause some discomfort in the intestine where they live, and generally cause no other more serious pathology, there are serious differences when we consider their larval cysticercus stage. *Taenia saginata* eggs when ingested by cattle will develop into cysticerci only in the muscles of cattle. This is because there are distinct differences in the type of bile salts secreted by herbivores (salts conjugated with the amino acid glycine) and those secreted by carnivores (conjugated with the amino acid taurine).

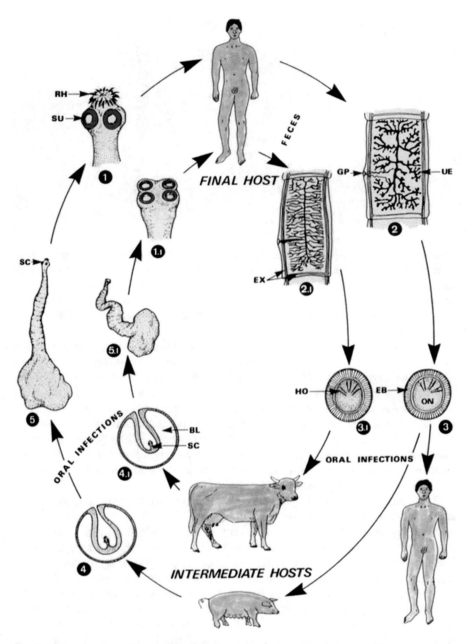

Fig. 3.4 Taenia solium (Source: Encyclopedia of Parasitology Ed. Melhorn 2016, Fig. 1)

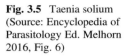

Fig. 3.5 Taenia solium
(Source: Encyclopedia of
Parasitology Ed. Melhorn
2016, Fig. 6)

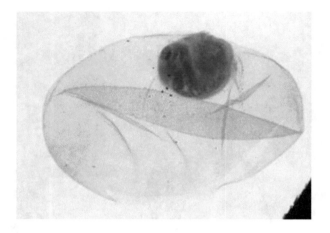

Thus the glycine based bile salts in cattle are thought to stimulate the oncospheres of *T. saginata* to develop into cysticerci in cattle and their absence in human bile salts makes humans unsusceptible to cysticercosis by *T. saginata*. Therefore when a person happens to ingest a cysticercus of *T. saginata* from undercooked beef in a rare steak or steak tartar, he or she can only acquire taeniasis and not the more dangerous cysticercosis.

In the early twentieth Century there was a spate of sensational newspaper stories in the United States about quacks promoting tapeworms, presumably *T. saginata* cycticercus larvae, as a way to lose weight in Midwestern cities like Peoria, Illinois. The narrative was sort of to keep a pet tapeworm inside you to share your food and keep your figure at the same time! This controversy went as far as then US Surgeon General Robert Blue who had to investigate the allegations and issued a denial in 1921 to debunk the stories. There were subsequent stories in Europe, likely apocryphal, that operatic diva Maria Callas herself tried the tapeworm diet. Thus the popularity of celebrity urban myths and parasite stories had always held a strange power to fascinate the general public.

Treatment for taeniasis (for the adult tapeworm) is with praziquatel or niclosamide as recommended by CDC.

However unlike *T. saginata*, the severity of human infections with *T. solium* is much more serious because humans can be infected by the cysticercus stage as well. The eggs of the pork tapeworm *T. solium* can develop in people and cause a serious infection called cysticercosis. Since humans are omnivores and have long domesticated its co-omnivore the pig, *T. solium* would have likely co-evolved with both the human and porcine intermediate hosts for thousands of years (Fig. 3.6). Humans may be infected by the cysticercus stage in several ways. Most commonly, uninfected persons acquire the infection from food handlers carrying the tapeworm in their gut, whose unhygienic habits resulted in their hands being contaminated by unsanitary toilet practices. Or individuals, who are infected with a tapeworm in their own gut, self-infect themselves with the parasite eggs by careless hygiene. There may be also a third possibility, although never proven, that the adult tapeworm in a

person's intestine may cause sufficient irritation to interfere with normal peristalsis and cause gravid proglottids to be transported by reverse peristalsis back into the stomach and thereby initiating the process that causes the eggs to hatch and develop into cysticerci.

No matter how it is acquired, human infection by the cysticercus stage of *T. solium* is serious as it may cause severe symptoms depending on where the cysts develop and their numbers, stage of development and size. They may develop in the muscles, eyes (Fig. 3.7) and also the brain and spinal cord (Fig. 3.8), and infections of these latter locations result in neuro-cysticercosis which is the most serious form of disease and one estimated to cause 50,000 deaths annually world-wide. Severe manifestations of neuro-cysticercosis include seizures and severe headaches, confusion, and difficulty with balance. Some parasitologists have hypothesized that Julius Caesar's well documented late onset epileptic seizures were in fact a result of neuro-cysticercosis. Julius Caesar (100–44 BCE) started having seizures when he was 54 years old, after returning from military campaigns in Egypt. During the Battle of Thapsus in what is modern day Tunisia, he suffered his first recorded episode of a history of seizures, subsequently followed up by many others including those during the Battle of Manda near Cordoba in Spain. A very interesting paper by R.S. McLachlan argued persuasively that Caesar might have likely acquired cysticercosis during his campaigns in North Africa and parts of the Middle East and what is present day Turkey, where there had been recorded prevalence of cysticercosis during those times. After he returned to Rome and was trying to consolidate power in the face of opposition from the Roman senators, he had further neurologic complications including severe headaches, dizziness and confusion, as well as seizures and loss of consciousness sometimes in the middle of senate debates. These episodes, which were used by his enemies to suggest weakness and deteriorating health, no doubt emboldened the political opposition, culminating in his historic assassination. Thus parasite infections such as cysticercosis probably played pivotal roles in human history in different parts of the world. There are other examples of parasites having an impact on the course of history, as we shall see later.

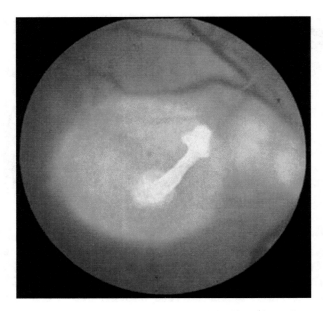

Fig. 3.7 Taenia solium (Source: Encyclopedia of Parasitology Ed. Melhorn 2016, Fig. 8)

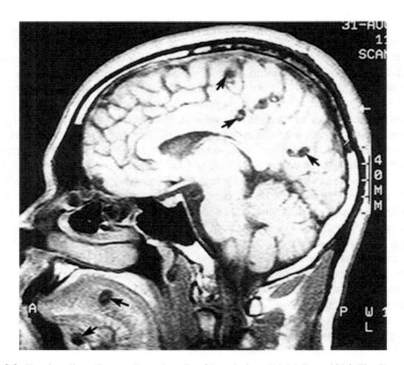

Fig. 3.8 Taenia solium (Source: Encyclopedia of Parasitology Ed. Melhorn 2016, Fig. 7)

Cysticercosis is prevalent in many low resource regions of Asia, Africa and the Americas, wherever the rearing of pigs occurs in areas lacking proper garbage disposal, sanitation, clean water and available access to medical care. Human carriers of the adult tapeworm of *T. solium* in their gut will be potential sources of infection to anyone with whom they come in contact. Commonly they are food handlers such as street food vendors, restaurant waiters, kitchen workers, as well as domestic help. The report in 1992 of an infection of four members of an Orthodox Jewish family in New York City who contracted cysticercosis was initially met with bewilderment because their religious practices precluded any contact with pigs. It turned out that their Mexican maid was a carrier of an intestinal *T. solium* tapeworm and was the source of infection, thus emphasizing the importance of the role of asymptomatic carriers.

Taeniasis due to *T. solium* and cysticercosis are starting to make a comeback in China since the opening up of their economy in 1989, and cysticercosis is now considered an increasingly serious public health problem. In the decades following the revolution of 1949, improved public health regulations, centralized government regulated slaughter houses and strict implementation of meat inspection resulted in drastic reduction in incidence. Since the 1989 economic reforms however, there have been a proliferation of uncontrolled small private butchers, backyard slaughter houses and a general deterioration in the quality of meat inspection. Economic reforms generally bring benefits by freeing up entrepreneurship and promoting prosperity in many countries in Asia but they also introduce hidden costs not usually reported in the popular media.

Thus when I first visited India on university exchange programs, I had expected to see cows freely roaming the streets because of their privileged holy status among the majority Hindu population there. I was however totally unprepared to encounter large groups of freely roaming pigs foraging among piles of garbage when our car reached rural areas in the outskirts of Jaipur in Rajasthan. I learned that India in fact has an extensive pig rearing industry, and apart from Muslim areas, these may be an increasingly important part of the local economy in many states in India. Recent reports in Indian medical journals indicated that CT scans and MRI imaging revealed far greater incidence of neurocycticercosis than previously thought. Thus cysticercosis continues to remain an underestimated and underreported disease in various unexpected parts of the world.

In parts of Mexico and Ecuador, for example, incidence of cysticercosis acquired from street food is rising alarmingly since contamination is marked by the high prevalence of tapeworm carriers among food vendors. Furthermore, hygienic food preparation is complicated by lack of clean water sources for the street vendors. As I am an avid foodie myself, and one who finds a foreign trip incomplete if I did not sample local street food, this posed a real dilemma for me while I was driving through the Andean highlands of Ecuador, when I was visiting our academic collaborators at the Universidad San Francisco de Quito. My usual rule of thumb is that street food is relatively safe if I follow two cardinal rules. Rule No. 1: eat only well cooked food which are grilled, fried, boiled, and cooked in front of you. Rule No. 2: avoid cold drinks sold in the streets, fruit juices, ices and popsicles and cut fruits of any kind. These two rules have long held me in good stead, as I have avoided any foodborne infections or tummy upsets despite decades of travel and sampling street food in many parts of the world. However I become paranoid when I travel to places known to have high prevalence of *T. solium* and cysticercosis. In this case the source of infection is not what is intrinsically in the food, which would usually be killed by high cooking temperatures, but rather due to its contact with the food handler after cooking, and the risk of very serious disease far outweighs pleasures of the gourmand in me. Thus as we drove through the highlands from Quito to Mindo and passed food vendors tempting the inner Anthony Bourdain in me with sizzling barbecued "cuyi" from their stalls at the roadside, we resisted and drove on. Cuyis are large rabbit sized sylvatic guinea pigs belonging to the rodent family, raised for food in many parts of South America. My friend and colleague Ricardo Izurieta, who is Ecuadorean, introduced me to cuyis in their many forms later when they were prepared in hygienic conditions in restaurants. On that day however, we speeded on to Mindo and were rewarded with the spectacle of thousands of humming birds of various species, for which Mindo is justifiably famous, darting all around us in a secluded sunlit clearing in the middle of a forest, an idyllic setting complete with a sparkling mountain stream running through it.

Taenia asiatica is the third Taenia species that causes taeniasis. This species is morphologically almost identical to *T. saginata* but its intermediate host is the pig in various parts of Asia such as Taiwan, China, Japan, Korea, Vietnam, Indonesia, the Philippines, Laos, and Thailand. In pigs the cysticercus is found in the liver, and human infections are associated with eating raw or undercooked pork liver, not uncommon in many parts of Southeast Asia. Current clinical observations conclude that humans can only acquire taeniasis, that is infections by the adult tapeworm in the intestine, and cysticercosis does not occur. Some parasitologists believe that the jury is still out because current diagnostic methods would not be able to distinguish among the various species of the cysticercus stage and many cases of cysticercosis reported as caused to *T. solium* may in fact be those of *T. asiatica*. They furthermore argue that *T. asiatica* may not be geographically limited only to Asia for the same reason, and newer molecular techniques may yet reveal its spread to places outside Asia especially where large communities of immigrants live.

Treatment for cysticercosis includes albendazole and praziquantel as recommended by the CDC.

***

Several other minor tapeworm infections mainly from natural animal cycles may cause human cysticercosis, namely *T. crassiceps, T. ovis, T. hydatigena*, and *T. taeniaeformis*. These are classified as zoonotic infections, that is humans are incidental hosts inadvertently infected by these parasites, whose natural cycles involve animals. These infections are extremely rare and very sporadic however. Another group of larval cestode infections in humans are the coenurus infections and these include *T. multiceps, T. serialis* and *T. brauni*. The difference between cysticercus and coenurus cysts is that the former has only one protoscolex per cyst, whereas in coenurus cysts multiple protoscoleses are formed and some may be free floating in the fluid within the cyst. Coenurus infections however involve the central nervous system and infections which may involve the brain and eyes and that make these infections extremely serious.

When I returned to Malaysia after college in Australia, I worked at the University of Malaya in Kuala Lumpur initially as a tutor and then as a junior faculty member while concurrently working on my doctoral dissertation. "Eddy" F.Y. Liew whom I had met at ANU in Canberra, and who stayed at the same dorm, Burton Hall, was a fellow Malaysian a few years ahead of me and was then a newly hired Lecturer in immunology at the University of Malaya. He had trained as an immunologist at ANU's John Curtin School of Medical Research and he became my mentor and major professor. He was to later embark on a distinguished career in the UK becoming Professor and Gardiner Chair of Immunology at the University of Glasgow and elected Fellow of the Royal Society, the highest honor for a scientist in the UK.

The project I embarked on was to study the immunobiology of experimental infections of laboratory rats with *Taenia taeniaeformis* cysticercosis (Fig. 3.9). I had chosen this project because the field of immunology of metazoan parasites, the study of immune responses to large parasites such as the tapeworms, flukes and roundworms was just taking off then, and the opportunity of studying with a young brilliant immunologist like Eddy was too good to pass. Besides I was then newly married, had just started on my career in academia and was about to start a family, and I wanted to work on a laboratory based project in experimental parasitology which had a better chance of being completed in a defined period of time. I was itching to get the doctorate under my belt and then to venture out and explore the real world of parasites and parasitic infections then prevalent in the tropics of which Malaysia was a part. The fact that *T. taeniaeformis* was a cestode and therefore a familiar parasite similar to Spirometra which I had studied in Australia, seemed a sensible choice. I studied the humoral and cell mediated immune responses against *T. taeniaeformis* infections and showed that it was possible to experimentally immunize rats with purified fractions of antigens of the cysticercus stage, thus anticipating strategies for producing vaccines against metazoan parasites.

***

There are several other minor tapeworm infections in humans which are zoonoses that we acquire from animals which are the natural hosts. Sometimes the animal hosts are household pets such as dogs and cats which occasionally transmit their parasites to us. A common dog tapeworm *Dipylidium caninum*, which despite its name also infects cats, used to be a common zoonotic infection of children in developed countries including the United States until fairly recently (Fig. 3.10). The life cycle involves fleas and children usually acquire infection from the accidental ingestion of infected dog fleas, or if the dog happens to nip an infected flea with its teeth and lick the child's face or mouth it will introduce the cysticercoid larval stage that way. Widespread improvements of veterinary care of pets in treating for tapeworms and flea reduction are the main reasons for the decrease of

Fig. 3.9 Larvae of Taeniaeformis in rat liver. Picture by author

Fig. 3.10 Dipylidium caninum (Source: Encyclopedia of Parasitology Ed. Melhorn 2016, Fig. 4)

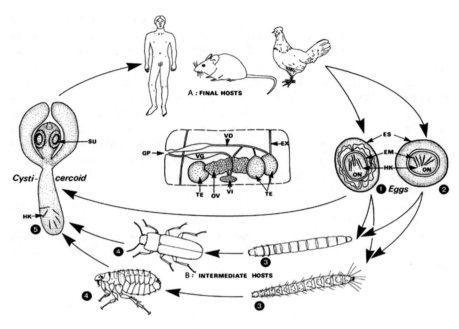

Fig. 3.11 Hymenolepidae (Source: Encyclopedia of Parasitology Ed. Melhorn 2016, Fig. 1)

dipylidiasis, although occasional cases are still seen. Treatment is with praziquantel and/or niclosamide.

In another tapeworm *Hymenolepis diminuta*, the intermediate host is the rat and transmission is via more than twenty species of insects such as fleas, flour beetles, moths and roaches which act as intermediate hosts (Fig. 3.11). Because *H. diminuta* is also called the Rat Tapeworm, it has acquired somewhat of a social stigma, being perceived as being associated with low rent neighborhoods. So it was a surprise and embarrassment when a group of socially eminent ladies who were members of a bridge club in toney Sarasota, Florida were found to be infected with *H. diminuta*. My friend the late Don Price who was an intrepid parasitologist who had done research in places as disparate as the Congo, India and Malaya, was then living in nearby Bradenton. Through a physician friend he was brought in as a consultant and when he confirmed that the stool samples were indeed positive for *H. diminuta,* he was asked to discreetly conduct research on how this might have occurred. In an investigation which would do Hercule Poirot proud, he determined that one of the ladies regularly brought a bag of "trail mix" cereal snacks for the weekly bridge games from an artisanal specialty store selling homemade cookies and other snacks. Cereal beetles of the species *Tribolium confusum*, a very common pest of cereal products, were found to be infected with *H. diminuta* cysticercoid larvae, and the beetles and their fecal matter had been contaminating the snacks. The multicolored pellets of odd sizes and shapes in trail mix no doubt made it easy to munch on fecal pellets while concentrating on concluding a rubber in the bridge games! No doubt that bridge club stopped buying from that particular store after that. The results of that investigation were never published but the mystery was quietly solved and no doubt the genteel ladies continued their bridge parties unruffled by any further untoward social scandal of this kind.

Hymenolepis nana is known as the Dwarf Tapeworm and is closely related to *H. diminuta*. They are unusual among cestodes in that they may entirely bypass the requirement of an intermediate host, although they can also infect intermediate insect hosts similar to those of *H. diminuta* such as fleas and flour beetles. However *H. nana* can also live completely only in the human host because when the eggs hatch, the cysticercoid larva can penetrate the villus of the intestines and encyst there. They can then re-emerge and grow into a new tapeworm in the same person's intestines, thus not only completing their entire life in a single host but also having the ability to be transmitted from person to person. Because it can be transmitted via the fecal-oral route, they commonly infect children, and inmates of psychiatric institutions. These infections cause diarrhea, abdominal pain, nausea and weight loss. Treatment for hymenolepiasis is with praziquantel, niclosamide or nitazoxanide.

Chapter 4
Trematodes: It Was Just a Fluke

I started my teaching career at the School of Biological Sciences, and later the Department of Zoology when the School was split into separate departments, at the University of Malaya (UM) in Kuala Lumpur in late 1971. I like teaching and interacting with young enquiring minds and immediately realized that I had somehow stumbled onto a dream job and that life in academia was the life I wanted. During those days, starting out as junior faculty member meant teaching large undergraduate classes of about a hundred, but I was fortunate also to teach the elite "Honors" class usually of just ten to twelve students who had been rigorously selected to complete their final year by doing research for an Honors Thesis. In a developing country like Malaysia in the 1970s, which had only two universities then, the other being the newly formed Penang University (since renamed Universiti Sains Malaysia or USM), it meant that the students in the Honors classes were indeed the *crème de la crème* of the entire country. It was therefore a privilege to mentor these very smart and eager students in their research projects.

It also meant that I had to design my own courses and organize laboratory classes to make the classes challenging. When I first joined UM many of the classes relied on teaching out of textbooks and using prepared slides and preserved specimens from the teaching repository, ensuring that the classes were incredibly boring. Coming from JD Smyth's department at ANU, I was determined that my classes would involve groups of students working with live parasite specimens such as examining the hatching of *Ascaris* eggs, infecting rats with trypanosomes and examining them in blood, making blood smears of rodent malaria, culturing hookworms to examine the living infectious larvae climbing up charcoal mounds in Petri dishes, microscopic examination of swimming microfilariae on a wet mount, and so on. I spent a lot of time visiting the local abattoir, coming back to the lab with buckets of live wriggling *Ascaris* worms from pigs, slithering dark brown leaves of liver flukes from cattle, and *Haemonchus* roundworms from goats.

It was during this period while I was visiting slaughter houses and local farms around Kuala Lumpur that I came to appreciate how integrated farming methods were practiced with great resource efficiency in SE Asia despite severe limitation of

© Springer International Publishing AG, part of Springer Nature 2017   35
B. H. Kwa, *The Parasite Chronicles*, https://doi.org/10.1007/978-3-319-74923-5_4

available arable land. Agriculture resembles high intensity gardening since the plots were very small and the farmers had to squeeze every bit of arable soil to make a living. In the early 1970s these farms on the outskirts of KL, as Kuala Lumpur is called by everyone who lives there, are sometimes just a couple of hectares in size. Many of these farms were run by tough hard working Hakka families. There would be some pigs corralled by crude fences of rough-hewn timber and sheltered under a rudimentary roof of *attap* palm fronds. We would usually find a small pond where ducks were reared and which also grew floating water plants which served as food for both the pigs and ducks. Often some other water plants were grown for human consumption and eaten as vegetables by the farmers as well. One ubiquitous example is the semi-aquatic "kangkong" (*Ipomoea aquatica*) which is nutritious and easy to grow on the bank surrounding the ponds. Pig excrement flowed from drains into the pond and acts as fertilizer for the water plants, and the pond itself would sustain fresh water carp. Because the pigs gained weight rapidly, have short gestation periods and large litter size, they were essentially cash machines and together with the fish and ducks, became valuable sources of income in efficiently managed farms. Fruit trees like papaya, banana, guava and the ubiquitous coconut in the tropics provided both food as well as additional cash crops. Vegetable plots surrounding the houses also contributed to the farmer's kitchen and/or household income and chickens provided food and eggs for the table. This was an efficient and sustainable system. The United Nations Food and Agriculture Organization (FAO) later promoted this type of tropical small scale farming as "Integrated Animal-Fish-Mixed Cropping Systems" and developmental economists hailed this as a pathway to reduce poverty in rural areas of the developing world.

This sustainable ecologically self-sufficient system is very vulnerable however and can easily be undermined by careless introduction of disease agents, and several parasitic flukes thrive very well within it. One such infection is the human intestinal fluke *Fasciolopsis buski*. This is the largest of the human flukes, or trematodes, and can grow to 7.5 cm or three inches in length. This is an incredibly beautiful parasite. Because the fluke is a flatworm, its organs are laid out flat like the intricate veins of a leaf. In parasitology classes, a large stained and mounted specimen is invariably demonstrated to students for the study of the morphological features of the flukes' digestive and reproductive systems, conveniently laid out flattened by nature for easy observation. The intricate patterns of the vitelline glands, involved in making the fluke's egg shell, and the branching testes and ovaries, look like the filigree prints of an inspired piece of Javanese batik fabric (Fig. 4.1).

The adult fluke is attached to the intestinal wall by a large oral sucker at its anterior end, and an acetabulum, sometimes called a ventral sucker, further down. Trematodes are hermaphroditic like the tapeworms described previously, and may reproduce both by self-fertilization or cross-fertilization, but mainly thought to be the former by most parasitologists. Fertilized eggs are released from the uterus of the fluke through its common genital pore and passed out in human feces. To survive, the eggs have to find themselves in an aquatic environment, as they are very susceptible to desiccation. Unfortunately, in many of the farms I visited, the farmers often build latrines directly over the ponds, presumably to add human excrement as

Fig. 4.1 Fasciolopsis buski
(Source: Encyclopedia of
Parasitology Ed. Melhorn
2016, Fig. 1)

fertilizer for the water plants (Fig. 4.2). This will thus ensure that the eggs will
survive and help to maintain several species of fluke infections including those of
Fasciolopsis buski.

Eggs of trematodes have an operculum and after sufficient time for embryonation
in water, they will be stimulated by light to hatch. The operculum will pop open and
release a free swimming ciliated stage known as a miracidium. The miracidium must
then find an appropriate snail intermediate host, penetrate its soft tissue where the
miracidia of *Fasciolopsis buski* will grow into sporocysts, within which several redia
larvae will form, and each redia in turn produce a second generation of even more
rediae. These intra-molluscan stages essentially serve to multiply enough progeny in
the snail host to ensure species survival, since very few miracidia would successfully
find and infect an appropriate species of snail in nature. All the major human
trematode parasites require a snail intermediate host and because they require at
least two hosts to complete their life cycle, are taxonomically named Digenean
flukes (Fig. 4.3).

From the redia the next stage, called the cercaria, will develop and leave the snail
and swim in water and those of *F. buski* will eventually attach to any available

Fig. 4.2 Fish pond next to latrine, pig farm outside Kuala Lumpur 1973. Picture by author

support such as the stems, fruits and leaves of water plants. Here the cercariae will form cysts, called metacercarial cysts, to protect themselves and will then lay waiting for its next host. Once ingested by a person, the metacercariae will excyst in the duodenum and attach themselves to the wall of the small intestines and grow into adult flukes. Although most infections are light and asymptomatic, heavy infestation especially in children can cause ulceration and hemorrhage at the attachment sites of the flukes on the intestinal wall, and they will also cause impaired vitamin B_{12} absorption, vomiting, nausea and anorexia. With implementation of public health measures and improved hygiene these infections are uncommon in Malaysia although they remain prevalent in many other parts of SE Asia.

In many parts of SE Asia such as Thailand, Vietnam, Laos, Cambodia and southern China, but less commonly in Malaysia nowadays, water chestnut plants (*Eliocharis tuberosa*), water caltrop (*Trapa natans*), and water lotus (*Nelumbo nucifera*) are grown in the farm ponds for human consumption. Metacercariae of *Fasciolopsis buski* are found encysted on the stems, leaves and fruits of these aquatic plants and while cooking them will kill the metacercariae, eating them raw will allow them to infect the human host. The fruit of the water chestnut (*Eliocharis tuberosa*) is crisp and crunchy and is often chopped up together with shrimp and pork to add texture and flavor to the little dumplings in "wonton" soup. The high cooking temperature of boiling soup would ensure that any metacercaria will be killed. Eaten raw they are sweet, crunchy and quite delicious, and children will use their teeth to bite and pull off the skin. If the skin is not washed carefully, they are probably a likely source of metacercaria in endemic countries. The water caltrop

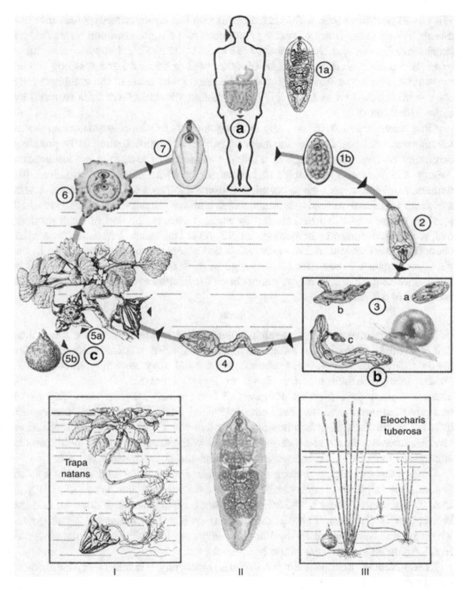

Fig. 4.3 *Fasciolopsis buski* (Source: Encyclopedia of Parasitology Ed. Melhorn 2016, Fig. 2)

(*Trapa natans*) with its hard brown skin and two horns resembling little water buffalo heads was a favorite snack delighting children in pre-industrial SE Asia with their natural "toy animal packaging", and I remember eating them boiled when I was a kid. Water lotus plants *(Nelumbo nucifera)* have green conical pods containing white seeds covered with thin green skin and these are always eaten raw as a snack.

We would peel the seeds with our teeth and pop the sweet crunchy seeds into our mouth. All of these fruits of water plants used to be quite common as traditional snacks in SE Asia and were probably the main source of *Faciolopsis buski* infections. Scrupulous washing of these fruits as well as careful hand washing would prevent infection, and that was probably the reason why none of the children in my class were infected when I was a kid even though we could have been exposed by eating these items.

On a recent trip in 2016 to Luang Prabang, a pretty city with exquisite temples in the highlands of Laos, I visited the morning market after participating in the predawn ceremony of making offerings to a silent procession of Buddhist novice monks. Among the fruits and vegetables that were displayed for sale by the vendors, the women resplendent in their colorful traditional clothes and head dresses, I was amazed to observe mounds of fresh water caltrops, bunches of lotus pods tied together at the stalks and baskets full of water chestnuts, the sort of fresh produce that were once common in markets in SE Asia but which I had not seen after economic development and modernization had swept through most of Asia in the past four decades. Prevalence rates as high as 30% of *F. buski* are still reported in Laos, thus indicating that improvements in public health have not yet reached all the countries in SE Asia.

***

There are numerous other intestinal flukes whose incidence rates are lower than that of F. buski. One of these is a group of several species of flukes belonging to the genus Echinostoma and other related genera, and they generally have a wide geographical distribution, from Japan to Egypt, Romania, Russia and Brazil. These are zoonotic infections with natural hosts which may be birds, dogs, cats and other mammals and humans being incidental hosts. For example, humans are infected by the following Echinostoma parasites when they ingest undercooked or raw intermediate hosts and become infected with larval stages of Echinostoma cinetorchis from frogs, E. lindoense from clams and E. japonicus from fish.

Other intestinal flukes belong to the genus *Heterophyes* and closely related genera. One important species is *Heterophyes heterophyes* that is found in Egypt, Iran, Tunisia and Turkey in the West and China, Japan, Korea, Indonesia, and the Philippines in the East. The other important species is *Metagonimus yokogawai* which is mainly prevalent in the Far East, although it has been reported as far as Israel, Spain and other parts of the Mediterranean.

Recommended treatment for fasciolopsis, heterophyiasis and metagonimiasis is with praziquantel.

***

Chinese New Year occurs around January and February every year depending on the Lunar calendar. One of the traditions, at least among the ethnic Chinese communities in SE Asia, mainly in Singapore, Malaysia, Hong Kong and Southern China, is the celebration of a dish eaten on Chinese New Year's Eve which is called "Yü shēng" in Mandarin. It is a giant bowl of salad with shredded vegetables, condiment dressing and thinly sliced slivers of raw fish, usually fresh locally farmed carp. As Chinese New Year is the Spring Festival, the entire family and guests would toss the salad together with chopsticks to symbolize rejuvenation and to augur prosperity in the coming year. The word "yü" for fish in Chinese is a homophone for prosperity or abundance, and that alone is usually reason enough to explain the popularity of one more festive Chinese activity which involves eating and family ritual during the New Year!

The carp unfortunately is an intermediate host of the Chinese liver fluke *Opisthorchis (Clonorchis) sinensis* and eating the fish raw is a risk for infection. Although this parasite correctly belongs to the genus *Opisthorchis*, it has been known as *Clonorchis sinensis* for such a long time that most parasitology books still use that name. The life cycle resembles that of *F. buski* but there are a few differences (Fig. 4.4). The intra-molluscan stages of *C. sinensis* have only a single generation of rediae, unlike the two in *F. buski*, but more importantly the cercariae that emerge from the snail intermediate host do not encyst on vegetation, but instead will seek out fish to infect. These are usually the freshwater carp reared in the ponds of the small farms mentioned earlier, such as the common carp (*Cyprinus carpio*), grass carp (*Ctenopharyngodon idellus*), and crucian carp (*Carassius carassius*). The cercaria will penetrate the skin of the fish, migrate to the muscles below the skin and encyst there as a metacercarial cyst. The metacercariae are very susceptible to temperature and will not survive common cooking temperatures. Therefore the infection cycle is maintained only when raw fish is consumed as in the "Yu Sheng" salad mentioned.

In the 1960s and 1970s there were sporadic reports of jaundice and liver complications after the Chinese New Year period in Malaysia and they were usually reported as being associated with *Clonorchis sinensis* infection of the liver. Obstruction of the bile ducts often resulted in jaundice, which is a diagnostic sign of probable liver fluke infection at that time in Malaysia. Although human liver fluke infections were never that widespread or important in Malaysia and Singapore, its seriousness cannot be overstated in other parts of Southeast Asia.

Ingested metacercariae exit from their cysts in the human host by a two-step process, the first event being when the outer cyst membrane is digested by host enzymes such as pepsin in the stomach. However the inner membrane only dissolves under a precise physiological regime in the host intestine consisting of temperature at 39° C, combined with CO_2, a suitable oxidation-reduction potential and the presence of bile. This second event was confirmed by *in vitro* culture of the trematodes, which allowed for elegant experiments of manipulation of individual physiological factors in culture media, and many of the *in vitro* methods for these parasites were pioneered by JD Smyth and his contemporaries. The metacercariae released in the intestinal lumen then enter into the opening in the intestine called the major duodenal papilla, through the ampulla of Vater (site where the common bile duct joins the pancreatic duct) and from there they migrate towards the many smaller bile ducts of the liver. There the metacercaria will develop and mature into the adult fluke within 4 weeks and start producing eggs.

Fertilized eggs which are passed by the mature flukes in the liver, enter the intestines from the bile ducts, and are then passed out in the stool. They will only

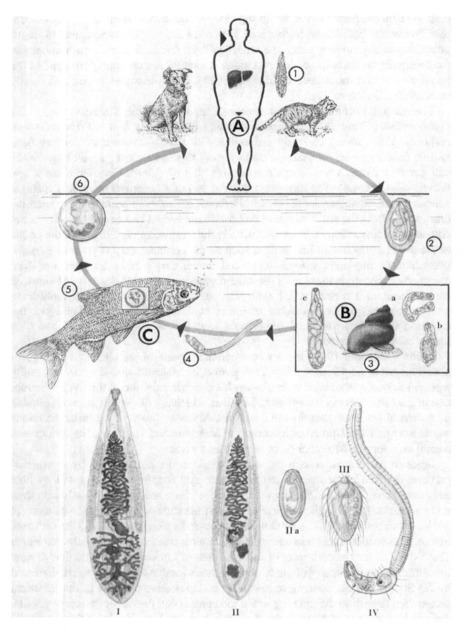

Fig. 4.4 Clonorchis sinensis (Source: Encyclopedia of Parasitology Ed. Melhorn 2016, Fig. 1)

survive in an aquatic environment, where they will embryonate and hatch to release the miracidia which will infect the snail intermediate host and begin their life cycle anew. Thus the *C. sinensis* fluke has evolved a life cycle that is perfectly adapted to the "Integrated Animal-Fish-Mixed Cropping Systems" that we see in the small farms of SE Asia.

As the adult flukes are relatively large, they will physically obstruct the narrow bile ducts within the liver and damage the surrounding liver tissue by mechanical pressure and irritation in part due to the action of the fluke's suckers. Additional pathological changes are induced by toxic metabolic excretions of the flukes, immunological reactions of the host, as well as secondary bacterial infection. The gall bladder often becomes non-functional, enlarged and containing viscous cloudy bile. Heavy infestations by *C. sinensis* worms have been associated with cirrhosis (fibrous degeneration of liver tissue), obstructive jaundice, pancreatitis (inflammation of the pancreas), cholangilitis (inflammation of the bile ducts) and even cholangiocarcinoma (cancer of the bile ducts). There is evidence that inflammation induced by the presence of these fulkes in the bile ducts may result in DNA damage and genetic alterations leading to neoplastic transformation. There appears to be a distinct pattern of gene mutations, chromosomal aberrations and epigenetic alterations in carcinomas of different etiologies associated with *C. sinensis* infection.

An interesting piece of archaeo-parasitological discovery came out of Cambridge University in England in 2016 when *researchers Hui-Yuan Yeh and Piers Mitchell* reported finding eggs of *Clonorchis sinensis* in tombs (Fig. 4.5) in a place called Xuanquanzhi near the famous archaeological discoveries in Dunhuang in northwest China. Xuanquanzhi was at the eastern end of the ancient Silk Road which connected the Chinese Empire to Europe and the Roman Empire around 130 BCE. The eggs were found on "hygiene sticks" which were sticks of bamboo covered in cloth which were the equivalent of toilet paper in those times. . . . . . . .yes, ouch! What is interesting is that the excavation site was in the arid Tamrin desert basin of northwest China, which did not have the aquatic habitats for *Clonorchis sinensis* transmission and the nearest suitable province was Guangdong in the humid sub-tropical region nearly 1000 miles to the south. Thus it showed that there had

Fig. 4.5 Clonorchis
sinensis (Source:
Encyclopedia of
Parasitology Ed. Melhorn
2016, Fig. 4)

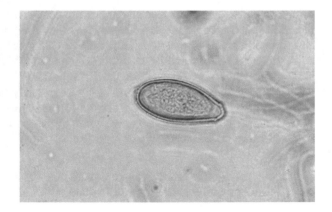

been a thriving trade route that connected southern China across Central Asia to Europe, and confirmed that the spread of infectious diseases along the Silk Road had accompanied the movement of people and goods. Globalization, with all its benefits and ills had therefore been around for several millennia and the current angst and debates about globalization are not unique to our times. As they say—*plus ça change, plus c'est la même chose.*

Opisthorchis viverrini is a closely related fluke, whose life cycle, target organs and pathogenesis are almost identical to *C. sinensis* but which can be distinguished by slight differences in their morphology and molecular structure. It is estimated that together they account for a total of 45 million infected persons worldwide with 35 million *C. sinensis* and 10 million *O. viverrini* infections. Their distribution is predominantly in East Asia, with the greatest prevalence in China, Thailand, North and South Korea, Laos, Vietnam and Cambodia. The incidence in Malaysia and Singapore has declined to a point that it is now mainly associated with travelers and those eating imported fish raw. A third liver fluke that occasionally infects humans is *Opisthorchis felineus* although the number of infections that are caused by this species is relatively small.

What makes human liver fluke infections of very serious concern is that chronic infections with both *O. viverrini* and *C. sinensis* are associated with very high rates of liver cancer in regions known to have high prevalence of liver fluke infections, mainly in Thailand, Korea, China, Taiwan and Hong Kong. In some parts of northeastern Thailand overall prevalence rates of opisthorchiasis mainly with *O. viverrini* reach an astounding 60%–80% and in one single community in Khon Kaeng, 100% of all residents over the age of 10 were infected! The liver cancer rates there are concurrently high as well. This was reported to be mainly a result of the popularity of eating chopped raw fish in a salad called "koi pla" and intense public health campaigns to promote cooking the fish before preparing the salad appear to have an impact as incidence rates among women and the young have dropped. The use of cooked fish in koi pla has increased within the household kitchens and fewer family members acquire *O. viverrini* that way. However the rates among the adult male population are still very high as koi pla made with the traditional chopped raw fish is a very popular dish accompanying drinking sessions that men participate in outside of the main meals, especially during the rice harvest season. There is also the general erroneous belief that the rice whiskey they drink will kill the parasites. Of course alcoholic drinks do not have any effect on fluke viability, and despite popular belief among many of the villagers, the condiments of salt, fish sauce, lime juice, hot chilies and herbs also do not kill the metacercariae in the fish. It is therefore unsurprising that most parasitologists I know, especially the ancient ones like me, avoid eating raw fish of any kind despite the contemporary popularity of eating raw fish among the young urban elite in most cosmopolitan cities!

Treatment is with albendazole or praziquantel.

***

There is at least one other genus of human liver fluke, *Fasciola,* that is of fairly widespread worldwide geographical distribution. There are two species, *Fasciola*

Fig. 4.6 Fasciola hepatica
(Source: Encyclopedia of
Parasitology Ed. Melhorn
2016, Fig. 3)

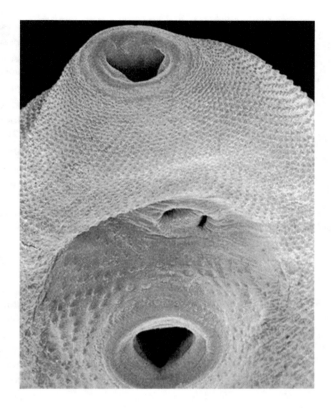

hepatica (Fig. 4.6) found mainly in temperate regions and *F. gigantica,* mainly in the tropics. Like the other human liver flukes, they involve snail intermediate hosts, and humans are mainly incidental hosts as the natural definitive hosts are usually sheep in the case of *F. hepatica*, and cattle, water buffalo, goats and other livestock in the case of *F. gigantica.* The metacercariae of these flukes encyst on vegetation where the natural hosts graze, but human infections usually occur with the incidental ingestion of vegetation infested with metacercarial cysts. In the case of *F. hepatica* in temperate regions such as in Europe, infections are commonly associated with eating infested water cress, *Nasturtium officinale*, in salads. Light infections are usually asymptomatic but heavy infestations can cause serious liver complications and ectopic infections are common. Occasionally unusual infections by young emergent flukes occur in the laryngo-pharyngeal region called "halzoun" in Lebanon and "marrara" in the Sudan are observed and these are associated with eating raw sheep liver.

It should be noted that finding Fasciola eggs in human stool is by itself not a conclusive diagnosis for fascioliasis, since spurious occurrence of the eggs in stool examination could be due to eating cooked infected liver which contained the parasite eggs. In this case the individual does not have an infection but is merely passing out dead fluke artifacts, and this type of so called pseudo-fascioliasis is very common in regions where sheep liver is a significant component of the local cuisine.

Dicrocoelium dendriticum is another liver fluke of herbivores such as cattle or sheep, and incidental infections can occur occasionally in humans. However *D. dendriticum* is famous for a totally different reason. When I was a graduate student, the story of how this parasite can control its intermediate host to enhance its own survival, was always held up as an elegant example of a rather ingenious manipulative strategy by some parasite species (Fig. 4.7). Eggs with the miracidia of *D. dendriticum* are passed in the feces of sheep or a similar herbivore and are eaten by snails. After two generations of sporocyst development in the snails, cercariae are released in slime balls secreted by the snails as they crawl on vegetation. Ants will eat the slime balls and become infected with the cercaria which then migrates in the ant's body and encyst as metacercaria usually in the hemocele of the ant. However those that encyst on the sub-esophageal nerve ganglion of the ant, which controls the ant's mandibles (pincer shaped mouth parts), will cause the ant to behave in a strange way that improves the parasite's survival and ensures the completion of its life cycle. After sundown when the temperature drops, the infected ant will climb up to the tip of a blade of grass and clamp down with its mandibles, and remain motionless all night. They remain like that through the dawn next morning and only release the mandibles and climb down when the sun is up and temperatures start to rise. To remain in that position during the high temperatures of the midday sun would probably kill the ant and the metacercariae within it. Since the peak grazing activities of the sheep and cattle definitive hosts are at dawn and in the evenings, this behavior of the infected ants ensures the best chance for the parasite to be ingested and to enter the final definitive site, the sheep or cattle intestines and eventually their liver. Although the actual mechanism of control of the ant's behavior is still not fully understood, this does confer a significant selective advantage for species survival of *D. dendriticum*. This is held up as a classic example of one of the ingenious ways which parasites have evolved to manipulate its host for its own benefit.

***

There is one more important fluke, which specifically infects the lungs of humans. In many countries in the East Asia, such as China, Japan, Korea, Taiwan, the Philippines, Thailand, Laos, Vietnam and Indonesia, but also in South Asia such as India and Sri Lanka, as well as island nations in the Pacific like Samoa and the Solomon Islands, the various culinary customs of eating raw or undercooked crustacea such as crabs, crayfish and shrimps exposes the inhabitants to the lung fluke *Paragonimus westermani*. These foods include in China "drunken crabs" (raw crabs pickled in rice wine overnight); in Thailand "kung ten" (raw shrimp salad) and "nam prik poo" (sauce made with raw crab meat and roe); in the Philippines "sinugba" (roasted crabs) and "kinilao" (raw crab meat); and in Korea "ke jang" (crab immersed in soy sauce). Other *Paragonimus* species are prevalent in Africa (*P. africanus*) and South America (*P. mexicanus*) as well, but their incidence rates are much lower.

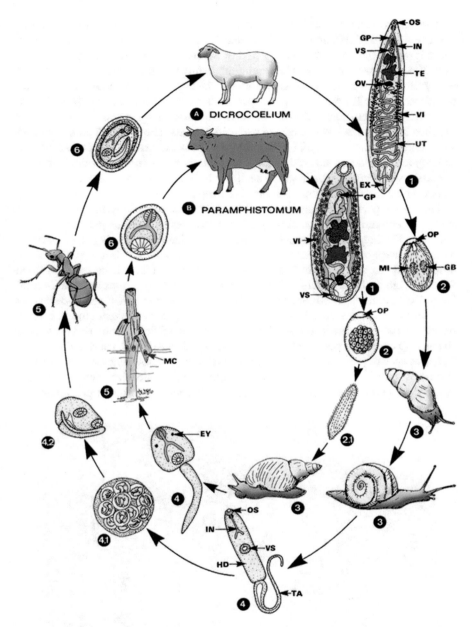

Fig. 4.7 Dicrocoelium dendriticum (Source: Encyclopedia of Parasitology Ed. Melhorn 2016, Fig. 1)

   The metacercaria encysted in the flesh of the crustaceans when ingested by people will excyst in the duodenum, and the metacercaria will penetrate the intestinal wall and migrate in the abdominal cavity first to the liver where they will grow into an adult fluke in about a week. The young adult fluke then re-enters the abdominal cavity from the liver and from there penetrates the diaphragm and enters into the lung tissue where it will be enclosed in a pseudocapsule and mature into adult flukes. There they will remain sometimes for years and produce eggs which will be either expelled in sputum or swallowed and excreted in feces. The eggs will embryonate in water and hatch into miracidia which will penetrate the flesh of aquatic snails and develop into sporocysts and rediae and complete their intramolluscan development. Cercariae will then emerge from the snail, swim in the water and seek a crayfish or crab to infect in order to complete their life cycle.
   Pulmonary paragonimiasis is a serious infection starting with milder symptoms similar to chronic bronchitis. However it can progress to serious pulmonary complications like hemoptysis, bronchopneumonia and lung abscess (Fig. 4.8). Furthermore, *P. westermani* fluke can sometime migrate into the brain and 25% of hospitalized cerebral cases in endemic areas in East Asia have been reported to be due to paragonimiasis. Cerebral infections used to be a very serious problem in rural South Korea until the 1980s but rapid and widespread industrialization has destroyed the habitats of the snails and freshwater crabs and have made them almost extinct. Ironically, the inadvertent consequence of environmental degradation and ecological changes (as well as better education about the disease) had caused the virtual disappearance of paragonimiasis in South Korea today. It however remains a serious disease in many SE Asian communities where local delicacies include raw and incompletely cooked crustacea. Treatment is with triclabendazole or praziquantel.

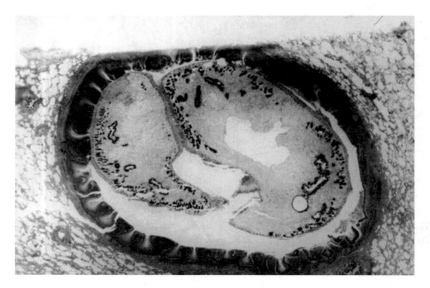

Fig. 4.8 Paragonimus species (Source: Encyclopedia of Parasitology Ed. Melhorn 2016, Fig. 4)

In my travels in SE Asia I always try to anticipate occasions when my local host would order a local delicacy which includes raw ingredients such as shrimp, crayfish, crabs, clams or raw fish of some kind and I will usually find polite ways to decline. So far I have managed to survive many late suppers and drinking sessions with local friends in roadside food stalls and night markets in far corners of the world and intend to continue to do so with a modicum of common sense, courtesy and street smarts.

Chapter 5
Schistosomiasis: Napoleon, Snails and Stalemate Across the Taiwan Straits

Parasitologists are very lucky that the ancient Egyptians practiced mummification. Think about it—apart from written records, we know very little about the epidemiology of diseases in ancient times: where they occurred, who were infected, how many were infected, were certain occupations at higher risk of infection, were men more likely to be infected than women, children more than adults? Moreover recorded histories were usually about kings and pharaohs, and they were written to glorify their conquests and victories, and precious stone tablets and papyrus were usually not wasted on recording the lives of ordinary farmers, household servants, carpenters and bakers. We can thus be grateful that the funerary practice of preserving the dead as mummies, among even the moderately wealthy inhabitants of ancient Egypt, provided objective records that answer many of the questions asked about at least one disease—schistosomiasis. Fortunately, the poor of ancient Egypt also bury their dead in shallow graves in the desert, where the high temperatures and dry air rapidly desiccated and preserved much of the corpses' organs and tissues by natural mummification. Therefore we can learn a lot about schistosomiasis in the entire social spectrum of society during ancient times.

Furthermore, there are two other factors in the nature of the schistosome parasite that provide us with an exclusive ability to accurately study the epidemiology of the disease in the ancient historical period, indeed even in the pre-historical period, since we are not dependent solely on written records. Firstly, because the schistosome parasite leaves behind eggs with hardened egg shells which survive as archaeological artifacts, we know the species of the parasite and type of disease that were prevalent during ancient times, as well as the extent of their spread among the human population. Secondly, because their intermediate host is the snail, and snails also leave behind a hard inert shell that remains intact for thousands of years, we can learn about the specific type, ecology, habitats and geographical distribution of the vector and the diseases that they spread with great precision. Therefore schistosomiasis provides us with unique archaeological fossils to chronicle an ancient disease accurately and objectively without relying solely on often biased, subjective and

© Springer International Publishing AG, part of Springer Nature 2017

B. H. Kwa, *The Parasite Chronicles*, https://doi.org/10.1007/978-3-319-74923-5_5

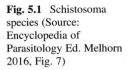

Fig. 5.1 Schistosoma species (Source: Encyclopedia of Parasitology Ed. Melhorn 2016, Fig. 7)

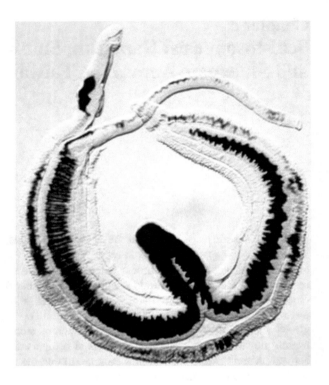

incomplete historical accounts. Indeed, we can use the fossils to verify the accuracy and veracity of the written records.

Schistosomiasis has been variously called snail fever, bilharzia, and bilharziasis, and the disease is caused by flukes of the genus *Schistosoma* that belong to the same family of trematodes that we had seen in the previous chapter (Fig. 5.1). There are several species worldwide: *Schistosoma mansoni*, the most widespread, is found in Africa, the Middle East, the islands of the Caribbean, Venezuela, Suriname and Brazil, having been brought over to the Americas from Africa with the slave trade; *S. guineensis* and *S. intercalatum* which are related, in the rain forests of central Africa; *S. japonicum* in China, Indonesia and the Philippines; *S. mekongi* in Cambodia and Laos. These live within the blood vessels of the human intestinal tract. Another species, *S. haematobium,* lives within the blood vessels surrounding the human urinary system and is found in Africa, the Middle East and the island of Corsica in the Mediterranean. Worldwide, the WHO estimates that schistosomiasis affects 240 million people in total, and another 700 million live in endemic areas. In sub-Saharan Africa alone it is estimated that 200,000 deaths per year are due to schistosomiasis.

Schistosomiasis is a waterborne disease and the schistosome parasite begins life when the egg, with the fully developed miracidium within the shell, leaves the human host via either urine in the case of *Schistosoma haematobium*, or feces in all the other species. The eggs have distinctive morphology which helps in their

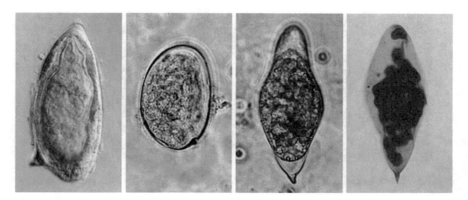

Fig. 5.2 Schistosomiasis, Man (Source: Encyclopedia of Parasitology Ed. Melhorn 2016, Fig. 5)

identification: those of *S. mansoni* have a large prominent lateral spine and are oval in shape; *S. haematobium* eggs tend to be long and narrow with a terminal spine; similarly the eggs of *S. intercalatum/S. guineensis* have a terminal spine but are even narrower in width; the eggs of *S. japonicum* are spherical with a tiny hook and those of *S. mekongi* are very similar (Fig. 5.2).

The eggs are prevented from hatching while still in the human host by the body's temperature and the osmotic pressure of body fluids, but once liberated into fresh water, *Schistosoma* eggs will be stimulated by sunlight to hatch and release a ciliated free swimming stage known as the miracidium. The miracidium must quickly find a suitable aquatic snail to infect as this actively swimming stage quickly depletes its energy reserves, usually within a day. Its penetration of snail tissues is assisted by two pairs of prominent glands which secrete powerful enzymes that allow it to enter the snail host. *Schistosoma mansoni* parasites will only infect snails that belong to the genus *Biomphalaria*; those of *S. haematobium* as well as *S. intercalatum* infect *Bulinus* snails; and those of *S. japonicum* infect *Oncomelania* snails; and *S. mekongi* the *Tricula* snails.

Within the snails the miracidium will develop two generations of sporocysts which serve to multiply the potential number of the infective cercaria stage that emerges streaming from the snail host. Each of the cercariae has a forked tail that allows them to be very motile in water, and is the crucial infective stage for the human host, as they will penetrate directly through human skin using its histolytic penetration glands Fig. 5.3). Therefore any kind of contact with water inhabited by infected snails will be a risk for infection with schistosomiasis. Once it reaches the subcutaneous tissues the cercaria loses its tail and the resulting schistosomulum will penetrate deeper until it finds a small blood vessel to enter. From there all the species of *Schistosoma* parasites except for *S. haematobium* will eventually reside in the mesenteric veins of the intestinal tract; those of *S. haematobium* will live in the venous plexus of the urinary bladder (Fig. 5.4).

The adult worms of schistosomes are dioecious, there are separate male and female worms. This is unlike all the rest of the trematodes, which are

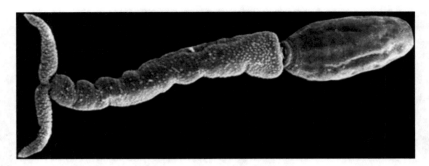

Fig. 5.3 Schistosoma species (Source: Encyclopedia of Parasitology Ed. Melhorn 2016, Fig. 3)

hermaphroditic, as we had seen in the previous chapter. The mature adult worms live their entire lives within the blood vessels *in copula*, in a perpetual reproductive embrace with the smaller and slimmer female worm held by the larger male worm within its gynaecophoric groove. Under the microscope the female worms appear much darker, as their uterus is densely packed with eggs which they release singly into the blood stream. The spines and hooks on the egg shells serve as anchors to prevent them being pushed back against the direction of the blood flow by back pressure. When the eggs eventually get pushed into the tiny venules and capillaries in the intestinal mucosa in the case of *S. mansoni*, or in the case of *S. haematobium* in the epithelial surface of the urinary bladder wall, the eggs secrete enzymes which allow them to escape into the lumen of the intestines or bladder. There the eggs are passed out in feces or urine, and are ready to hatch once in enters into water.

The various aspects of disease caused by schistosome infections are mainly a result of immunopathological reactions between the parasite and host immune responses. For example, one form of acute pathology is caused by the cercariae penetrating the skin, resulting in cercarial dermatitis due to IgE mediated hypersensitivity. Swimmers in the Great Lakes of North America will sometimes suffer cercarial dermatitis, called Swimmer's Itch, caused by cercariae of schistosome parasites of birds. These non-human parasites do not cause any severe disease other than dermatitis. Migrating schistosomula of the human parasites, however, may also cause an acute condition known as Katayama syndrome, similar to serum sickness, which is due to soluble immune complexes formed during heavy infestation, and which results in fever, fatigue, myalgia, malaise, headaches and pulmonary infiltration associated with a persistent cough.

However, the eggs of schistosomes cause the most serious chronic pathology associated with schistosomiasis. In the case of *S. mansoni* and *S. japonicum* this occurs in the walls of the large and small intestines respectively. Due to the intense production of eggs, the accumulated deposition of parasite eggs in the mucosa of the intestinal walls will lead to granulomatous reactions around the embedded eggs, causing ulcerative inflammation and intense formation of pseudo-polyps and bleeding in the intestinal tissues. As the mesenteries connect the intestines to the liver via the portal system, parasite eggs will also damage the liver.

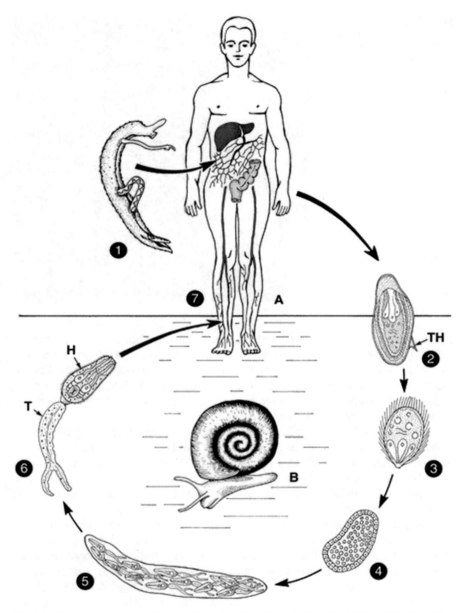

Fig. 5.4 Schistosoma species (Source: Encyclopedia of Parasitology Ed. Melhorn 2016, Fig. 5)

In the liver, a similar process due to embedded eggs in the hepatic sinusoids will lead to the development of granulomas and hepatomegaly (Fig. 5.5). The chronic infection and granulomatous reaction to soluble antigens released by the eggs will result in excessive collagen deposition in the hepatic matrix producing periportal

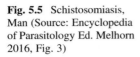
Fig. 5.5 Schistosomiasis,
Man (Source: Encyclopedia
of Parasitology Ed. Melhorn
2016, Fig. 3)

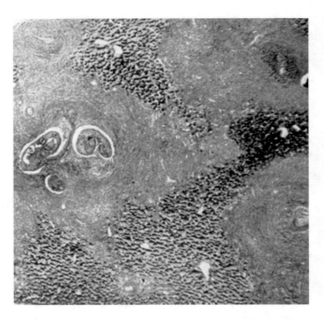

fibrosis and progressive occlusion of the portal veins. Eventually, the occlusion of the portal veins will lead to the development of portal hypertension, portacaval shunting, splenomegaly, gastrointestinal varices, ascites, and gastrointestinal bleeding which will ultimately be fatal. The chronic bleeding also results in serious anemia, which among adults leads to wasting, and among children results in growth retardation and cognitive impairment.

In the case of *S. haematobium* which lives in the veins surrounding the genito-urinary system, the eggs will be deposited in the wall of the urinary bladder and the ureters. Eventually fibrosis and calcification will occur and lead to secondary bacterial and fungal infections. In severe cases, chronic infection may result in renal failure. Furthermore in Africa and the Middle East, *S. haematobium* infection is reported to be associated with 50% of the bladder cancer cases, adding substantially to the total disease burden. In fact, *S. haematobium* caused such widespread prevalence of hematuria in many parts of Egypt in the past that the occurrence of so-called "male menstruation" had been accepted as a normal phenomenon, since most boys would have started to pass bloody urine by their teenage years (Fig. 5.6). Treatment of schistosomiasis of all species is successful with praziquantel.

In ancient Egypt, where this chapter began, fossilized shells of the snail *Bulinus truncatus* the main vector of *S. haematobium,* had been discovered in Late Palaeolithic sites at Edfu and Esna in Upper Egypt near the head waters of the Nile, indicating that conditions optimal for transmission of schistosomiasis must have existed in that area almost 40,000 years ago. However, it was the advent of large scale agricultural development and irrigation schemes in Egypt that probably

Fig. 5.6 Schistosoma
species (Source:
Encyclopedia of
Parasitology Ed. Melhorn
2016, Fig. 4)

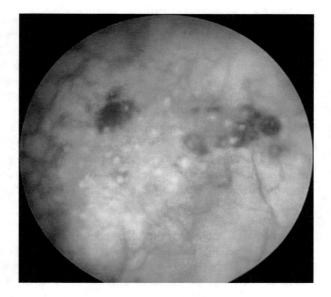

brought about the sudden rise in schistosomiasis infection rates, and this was
borne out by the extensive *Bulinus* fossil records that exploded around the period
from 4000 BCE onwards. Since houses then were built using mud bricks made
from the alluvial soil from the river banks, fossilized *Bulinus truncatus* snail shells
have been found in the bricks of the houses in excavated settlements along the
banks of the Nile and its tributaries. Similar *Bulinus* fossils have been found in
places outside Egypt such as in ancient Mesopotamia (present day Iraq) starting
from the period around 4000 BCE and in Jericho (in the present day Palestinian
West Bank) dated 1650 BCE, showing that *S. haematobium* was prevalent
throughout the Eastern Mediterranean region during the period that coincided
with the development of agricultural societies in the Western World. The construction of canals and ponds for agriculture created man-made snail habitats on a
large scale, and this intensified water contact by the increasing populations who
migrated to the area to catch fish and harvest shell fish; grow two kinds of wheat
(spelt and emmer) and barley to make bread and beer; grow flax to spin into linen
for clothing; harvested papyrus; and grew vegetables and fruits to supplement
their diet. Domestic daily water contact by activities such as washing clothes,
carrying water for household use, and bathing will further contribute to potential
infection by the waterborne cercariae penetrating the skin. All these agricultural
activities necessitated intensified and sustained contact with water, and as the
population increased, the lack of sanitation will easily result in an ideal habitat for
S. haematobium and its *Bulinus* snail host to spread.

Studies of mummies from 5000 years ago have revealed that schistosomiasis was widespread among the different age groups, gender and socio-economic status in Egyptian society. In the elaborate process of mummification, the lungs, stomach, intestines, liver, bladder, spleen and pancreas were all removed and placed in special canopic jars next to the body. The brain tissue was removed with a special hook via the nasal passage. Only the heart (and sometimes the kidneys as well) were left intact within the visceral cavity, which was then embalmed with natron (mainly sodium chloride and sodium carbonate). The contents of the canopic jars were similarly treated with the natron and then the body was sometimes filled out with sand and linen to preserve its shape and everything was allowed to desiccate in the heat in special chambers for 70 days. A fascinating book "Studies in the palaeopathology of Egypt" written by Sir Marc Ruffus, described the state of the organs and tissues of the mummies, as well as their associated diseases in great detail.

Mummies of adolescent as well as adult male mummies had been reported with *S. haematobium* eggs embedded in their mummified organs, indicating schistosome infection. There were also several occurrences of finding calcified bladders in some mummies which were indicative of long term chronic *S. haematobium* infection. Some naturally mummified corpses had been recovered from shallow graves, apparently from those lower on the social ladder, and identified by funereal clothing and buried artifacts as having had occupations as weavers, farmers and stone masons, and still others who were papyrus reed carriers and boatmen. Many of the wealthy mummified corpses, as expected, were from the landed gentry. One particular study of mummies from a village called Wadi Halfa was of particular epidemiological interest as it provided information on age and sex distribution: 7 males and 7 females, all between the ages of 15 to 40 were infected with *S. haematobium*. This confirmed that the age/sex distribution had remained quite stable since they are similar to the distribution of schistosomiasis prevalent in Egypt today. Thus researchers have been able to reconstruct a fairly accurate picture of the epidemiology of schistosomiasis in conditions that prevailed several millennia ago in Egypt.

An interesting debate had been going on for a long time regarding the controversy surrounding the death of Napoleon Bonaparte and the determination of the cause of his death. Most historians had concluded that he probably died of stomach cancer particularly as his father Charles Bonaparte had also died of stomach cancer. That familial connection, as well as the detailed pathological examinations consistent with stomach cancer, performed and reported by the pathologist and surgeon F. Antommarchi who performed the original autopsy, tended to favor this conclusion. Later studies by the Swedish dentist Sten Forshuvud among others, hypothesized that he might have been poisoned with arsenic, likely as a form of political assassination to ensure that there would not be another escape similar to the one from Elba. Subsequent analysis of Napoleon's hair found concentrations of arsenic, lending more credibility to the arsenic poisoning theory, although the post mortem account of Antommarchi did not provide any evidence that supported it. A third hypothesis, which was based on the descriptions of Napoleon's fibrotic liver pathology, as well as lesions in his calcified urinary bladder by Antommarchi, were highly suggestive of schistosome infection that likely could have occurred during his campaigns in Egypt which had lasted almost two years. This suggestion was considered plausible as the prevalence of schistosomiasis was very high in Egypt at that time, and it was corroborated by well documented accounts by French military physicians that many of Napoleon's soldiers under his command were infected by both *S. haematobium* and *S. mansoni*. We may never know, since microscopic examination for schistosome eggs were not described by Antommarchi in his autopsy report. Perhaps Napoleon had died of stomach cancer after all, but could have been concurrently infected with schistosomiasis that had hastened his death in St. Helena.

In the Far East, the practice of mummification was not developed to the extent in ancient Egypt, but early Chinese aristocratic families also sometimes preserved their bodies as mummified corpses. One such female mummy from around 2100 years ago from Hunan province was identified to be infected by *S. japonicum*. There have also been others in Hubei of a similar lineage. However the most interesting connection of schistosomiasis to an historical event in Asia happened in much more modern times. During the final period of the Chinese civil war between the Chinese Red Army under Chairman Mao Zedong and the Kuomintang (KMT) army commanded by Generalissimo Chiang Kai-shek, the defeated KMT troops were trapped in the southeastern provinces of China and were forced to make a dangerous evacuation across the straits to the island of Taiwan in late 1949. By the time Mao's troops had regrouped for the final battles in the south to capture the retreating remnants of the KMT army, most of them had successfully reached Taiwan Island using whatever cargo ships, fishing boats and coastal maritime vessels they could commandeer. Chiang Kai-shek had been well established in his new power base in Taiwan by then, and a final attempt by Mao's troops to capture the island of Quemoy off the coast of the mainland was thwarted at the Battle of Kuningtou. So Mao's Red Army was forced to camp out in the southern provinces to make preparations and practice for an amphibious assault on the island. They conducted their training and manoeuvers for cross-strait marine operations and amphibious landings in the lakes, basins and rivers in the southern provinces and had planned to launch the invasion across the Taiwan Straits at the appropriate time. Unfortunately for the Communist troops, the waters in the southern provinces of China were infested with *Oncomelania* snails carrying S. *japonicum* parasites and it resulted in one of the most devastating epidemics of schistosomiasis ever recorded, as an estimated 50,000 Chinese soldiers came down with acute *S. japonicum* infections (Fig. 5.7). By the time the Red Army had regained operational strength with fresh troops and consolidated their forces for the amphibious assault it was already too late. This was then early 1950 and the Cold War had intensified and the unstable division of the Korean Peninsula, with a Soviet backed North and a US backed South, had descended into sporadic hostilities and by June 1950 the full fledge Korean War had broken out.

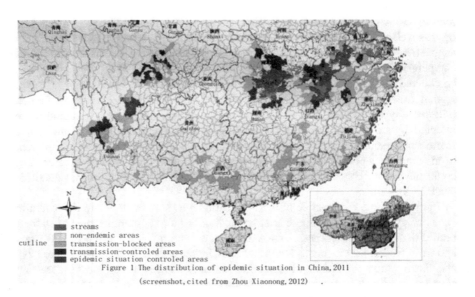

cutline
streams
non-endemic areas
transmission-blocked areas
transmission-controled areas
epidemic situation controled areas
Figure 1 The distribution of epidemic situation in China, 2011

(screenshot, cited from Zhou Xiaonong, 2012)

Fig. 5.7   Schistosoma japonicum (Source: Encyclopedia of Parasitology Ed. Melhorn 2016, Fig. 1)

Mao's army had to be re-diverted to the north, and the standoff in the Taiwan Straits between Taiwan and the People's Republic of China remains to this day, with the civil war technically unresolved. Thus one more parasite had played its pivotal role as *Deus ex machina* in the modern history of Asia!

Chapter 6
Filariasis and Elephantiasis: You Can't Go Home Again

There is a small town—actually it was then little more than a large village—called Pondok Tanjung in the state of Perak in Peninsular Malaysia. In the late 1970s it was a one street town fringed by orderly rows of rubber trees which reached out to the surrounding jungle, but even then the town already had a desultory look about it as production of natural rubber had slowly started to decline under the global onslaught of developments in the synthetic rubber industry (Fig. 6.1). The Malaysian economy by then was moving away from relying exclusively on primary production of commodities such as rubber, palm oil and mining and turning towards manufacturing in light industries like apparel, electronics and household goods. Pondok Tanjung was therefore in transition from a boomtown in the 1960s when commodity prices were high, and becoming an abandoned ghost town bypassed by new industrial parks in the outskirts of the larger towns in Perak State like Taiping and Ipoh, which had better infrastructure and transportation to the major export oriented ports. I was born and raised in Taiping just hardly 20 km away, yet I had not known about Pondok Tanjung until after I had started to do research in lymphatic filariasis. In one of the great ironies of my life, it turned out that Pondok Tanjung had the dubious distinction of having the highest prevalence of lymphatic filariasis in the country during that period in the 1970s.

Most people have heard of lymphatic filariasis because it is associated with a condition known as elephantiasis, whereby infected limbs and genitalia become grossly distended and enlarged and the skin thickened and roughened so that they resemble those of elephants, at least in the popular imagination. Although fully developed elephantiasis is only presented among a small proportion (less than 2%) of those with the disease, when left untreated over a period of years, its dramatic appearance is unforgettable to the observer (Fig. 6.2). Despite the fact that the hot spot for filariasis was essentially just in the outskirts of the small town of Taiping where I lived until I left for college, I had never seen a single person with elephantiasis while I lived there. The socio-cultural reasons that made a disease, with such grotesquely manifest clinical presentations, completely invisible will be explained later.

© Springer International Publishing AG, part of Springer Nature 2017
B. H. Kwa, *The Parasite Chronicles*, https://doi.org/10.1007/978-3-319-74923-5_6

Fig. 6.1 Outskirts of Pondok Tanjung, Malaysia, circa 1975. Picture by author

When I completed my doctoral dissertation and graduated with my PhD from the University of Malaya, I met Mak Joon Wah who was then a young physician and biomedical scientist who headed the Division of Filariasis and Malaria at the Institute for Medical Research (IMR) in Kuala Lumpur. At that time the IMR was laid out in a spacious campus of mainly wooden colonial style bungalows with wide airy verandahs that ran around each of the buildings, connected to each other by roofed walkways across broad lawns dotted with palm trees and hibiscus bushes. Belying this park-like setting amid the tropical torpor, Mak was a whirlwind of indefatigable drive and energy, with an incisive piercing scientific mind (Fig. 6.3). It was my good fortune to have had the opportunity to work with him, for he was a generous, good humored and friendly colleague and our mutual love of good food made the working relationship really enjoyable. The IMR was across town, through the busy and traffic jammed streets of downtown Kuala Lumpur, from my office at the university in the suburbs. However, Mak's encyclopedic knowledge of the best little gems of foodie *kopitiams* made my weekly commutes to his laboratory highly anticipated by me, for our research sessions would be capped by wonderful lunches in KL's many exquisite eating establishments minutes from the IMR. He was to become the Director of the IMR and later Pro Vice Chancellor of the International Medical University in Kuala Lumpur after I had left for the US.

Lymphatic filariasis prevalent at that time in Malaysia was mainly due to infections by sub-periodic *Brugia malayi*. The disease is caused by a mosquito-transmitted nematode worm which parasitizes the lymph nodes and lymphatic vessels of humans. Infections occur when infectious mosquitos bite a person and

Fig. 6.2 Tropical
Elephantiasis (Source:
Encyclopedia of
Parasitology Ed. Melhorn
2016, Fig. 1)

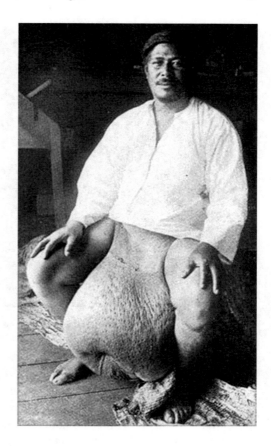

the infective third stage larvae of the parasite migrate from the mosquito proboscis into the bite wound on the skin. They then migrate from the sub-cutaneous tissue into the small lymph vessels where they molt, discard their outer cuticle and grow a new one. They thus become fourth stage larvae which will continue to migrate towards the larger lymph vessels and lymph nodes where they will undergo a final molt and develop into either male or female worms (Fig. 6.4).

The developing worms live in the lumen of the larger lymph vessels, as well as in the spaces within the lymph nodes through which the lymph flows. As they grow in size and excrete metabolic products which are highly antigenic, they will provoke a progressive inflammatory response from the host immune system. Chronic inflammatory reactions include intense infiltration of host immune cells, fibrosis, thickening of the vessels and surrounding tissues, leading to obstruction of the affected lymphatic circulation. Dilation of the lymph vessels cause them to balloon out as the normal flow of lymph becomes blocked. Externally, signs of inflammation in the lymphatics present as lymphangitis and lymphadenitis. The pitting edema accompanying infected limbs have a distinct characteristic of remaining depressed for a long time if you press down and make an indentation on the skin. The very thick and

Fig. 6.3 Mak Joon Wah during filariasis survey, Pondok Tanjung, Malaysia circa 1975. Picture by author

viscous lymph that oozes into the surrounding tissue from infected lymphatic vessels causes this type of lymphedema (Fig. 6.5).

This maturation of the worms may take up to 1 year, and they will then mate and the fertilized embryos within the uterus of the female worms will develop into microfilariae. These microscopic wormlike microfilariae are then released by the female worms and they will find their way from the lymph vessels into the blood capillaries connected to them, and from there into the general blood circulation (Fig. 6.6). This incessant daily release of thousands of microfilariae by the female worms acts as recurring antigenic challenges to the host immune system. If left untreated, the accumulated damage to lymphatic tissues will ultimately manifest itself as elephantiasis, especially when bacterial and fungal infections are superimposed on the compromised lymphatic functions. Elephantiasis caused by *Brugia malayi* usually involves the lower limbs, although those of *Wuchereria bancrofti* will involve the genitalia as well as the limbs.

Some filarial species have microfilariae that are enclosed by a sheath, which is a membrane analogous to the egg shell protecting the embryos of other nematodes for survival outside the host. Since filarial parasites have adapted to live in an insect vector, a protective egg shell is unnecessary and the sheath is an evolutionary vestige which is all that remains of the shell. The presence or absence of the sheath around the microfilaria is however very helpful to the parasitologist, as they allow us to distinguish between different species of the filarial parasite family. Thus, for example, *Brugia malayi*, *Wuchereria bancrofti* and *Loa loa* microfilariae are sheathed

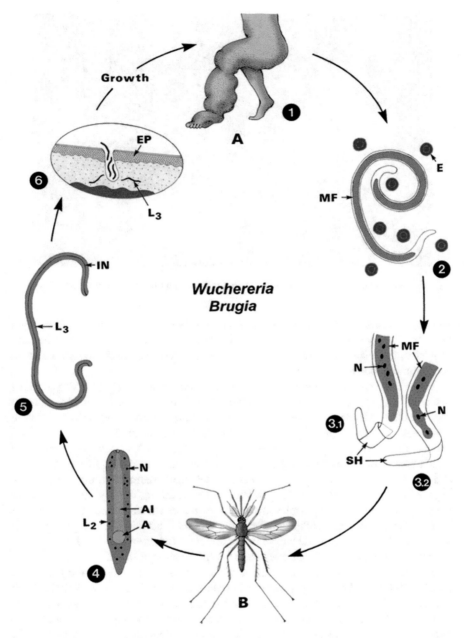

Fig. 6.4 *Wuchereria bancrofti* (Source: Encyclopedia of Parasitology Ed. Melhorn 2016, Fig. 1)

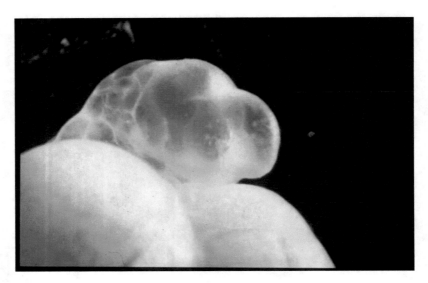

Fig. 6.5 Dilated nude mouse lymph node infected with Brugia malayi. Picture by author

whereas *Mansonella ozzardi*, *M. perstans* and *Onchocerca volvulus* microfilariae are not. This is very useful for their laboratory diagnosis in blood examinations in parts of Africa, for instance, where several of these parasites may be present in the same geographical location.

When they are in the blood circulation, the microfilariae exhibit a very curious behavior. They will appear in the peripheral blood from which doctors, by taking a drop of blood sample, may be able to observe their presence and make a positive diagnosis. However, the appearance of the microfilariae in the peripheral blood occurs only periodically. Some species and strains exhibit nocturnal periodicity which means that they only appear in peripheral blood in the middle of the night and are absent during the day. These are the nocturnally periodic strains. Others will appear in greater numbers at night but are still seen during the day, although fewer are observed when blood samples are examined under a microscope during daylight hours. These types are called sub-periodic strains. In Malaysia the predominant species in the 1970s was the sub-periodic strain of *Brugia malayi* and most of the efforts to control and eradicate the disease were directed at them.

So what happens to the microfilariae when they are not found in the peripheral blood circulation? They mainly migrate into the capillaries surrounding the alveoli in the lungs, where they will concentrate in the highly oxygenated pulmonary blood and only return to the general blood circulation during the designated time. This circadian regime is dependent on the host's physiological biorhythm during the course of the day, determined by the person's sleeping and waking activity patterns. It was observed for instance that in infected persons who travel through different time zones, their microfilariae will follow the time shift in their periodicity. The periodicity of microfilariae had apparently evolved to maximize the chances of the

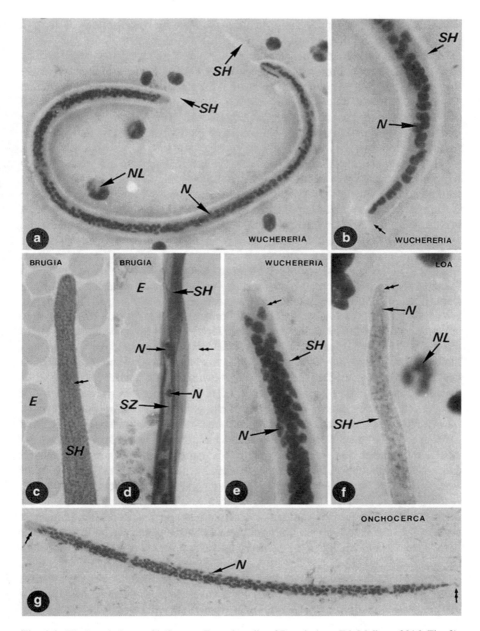

Fig. 6.6 Wuchereria bancrofti (Source: Encyclopedia of Parasitology Ed. Melhorn 2016, Fig. 2)

microfilariae being taken up by the specific biting behavior of the predominant species of mosquito vector in that part of the country. Thus if the mosquito species in that geographical location tend to only bite at night, those filarial parasite strains whose nocturnal periodicity are synchronized with the mosquitos will have a survival advantage.

A mosquito which bites a person carrying the parasite picks up microfilariae in its blood meal and the microfilariae will escape from the mosquito's midgut, penetrate the gut wall and migrate to the mosquito's thoracic muscles and develop into the first stage larva. They will continue to grow and molt twice, passing from the second stage larva to become eventually third stage larvae. This is the end point for the parasites in the mosquito, as they cannot develop any further unless they enter the mammalian host.

These third stage larvae are called infective larvae since this is the stage which will infect people. Some of them migrate into the mosquito proboscis which is the feeding tube which consists of the modified mouth parts of the mosquito, and there the larvae will line up waiting like arrows in a quiver. When the mosquito bites a person, these infective larvae migrate into the microscopic bite wound rapidly and resume the cycle in the human host.

The vectors of sub-periodic *Brugia malayi* in Peninsular Malaysia are mosquitoes belonging to the genus *Mansonia*. These mosquitoes breed in the riverine-swamp forest ecotypes of the coastal plains where the major rivers meander through the tropical forests after rushing down from the central mountain range. The larvae of *Mansonia* mosquitoes live among the roots of floating water weeds which grow in the jungle streams and irrigation canals which fringe the rubber plantations in states like Perak. Unlike most mosquito larvae which cling to the water surface to obtain their air supply, *Mansonia* larvae are adapted to obtain oxygen from the roots of the water weeds by specialized breathing tubes which attach to the fine rootlets of aquatic plants. The most common species of the vector mosquitoes for sub-periodic *Brugia malayi* in that part of the country were *Mansonia bonneae*, *M. dives*, *M. uniformis* and *M. indiana*, and because of their unusual microhabitat beneath the floating vegetation, conventional methods to control mosquito larvae by pesticide treatment was very challenging. This was especially true in the irrigation canals beside the rubber plantations, where fertilizer runoffs promoted the growth of aquatic weeds. When I visited the plantations with significant incidence of filariasis at that time, I observed that the floating vegetation in the canals were sometimes so dense that the water surface could not be seen. It was apparent that the solid mass of water weeds prevented spraying of liquid insecticides in places where it would have been impossible to reach the *Mansonia* larvae among the roots almost a foot below the water surface (Fig. 6.7).

Furthermore, the rubber plantations were mostly developed in clearings adjacent to the jungle and leaf monkeys such as *Presbytis melanophos* and *P. cristata* from the jungle would frequently forage among the rubber trees to feed on the nuts (Fig. 6.8). Since the leaf monkeys are the natural hosts of the filarial parasites, they act as reservoir hosts and maintain the cycle of infection among the population of plantation workers who live in the same area. Latex from the rubber trees flows best in the early dawn and the rubber tappers would be out among the trees between 4 and 6 am, coinciding with the peak biting times of the *Mansonia* mosquitoes which are mainly exophilic (living and resting outdoors) and exophagic (outdoor biters).

Fig. 6.7 Canal adjacent to plantation overgrown with aquatic weeds, Pondok Tanjung, Malaysia circa 1975. Picture by author

Fig. 6.8 Presbytis monkey in rubber plantation, Pondok Tanjung, Malaysia circa 1975. Picture by author

These same mosquitoes would feed both on humans as well as the monkeys, making filarial transmission between monkeys and humans almost impossible to break.

These challenges that faced the public health workers at that time meant that they could not depend on conventional vector control measures, removing one of the most cost effective tools available. The use of herbicides to kill the weeds which harbor the *Mansonia* larvae was difficult because it was expensive to cover the widely dispersed streams and canals in the fringes of jungle, and the frequent tropical rains would have diluted the efficacy of the herbicides. This made it necessary to concentrate resources on controlling the transmission of the disease in the human population using drug treatment. The most cost effective pharmaceutical drug then available against lymphatic filariasis was diethyl-carbamazine citrate (DEC) which many dog owners would know as Filaribits (given once a day) before they were displaced by the more convenient drug ivermectin (given once a week), sold as Heartgard in the United States. These were developed originally for veterinary use in the cattle industry to control for cattle nematodes and ectoparasites, and later reformulated for pets when it was found to be very effective for the dog heartworm *Dirofilaria immitis* which also belongs to the large family of filarial parasites.

It is a clear indication of the "orphan" status of both lymphatic filariasis and onchocerciasis, which we will discuss later in the next chapter, that the only available medications for millions of people suffering from these diseases worldwide were essentially developed for animals. It is a compelling indictment of the *status quo* in global economics that the pharmaceutical industry finds it more lucrative to cater to a market of medications for animal pets in North America and Europe than for humans in the poorer countries. However, in an amazing act of global citizenship, one company—Merck—donated its product ivermectin, rebranded as Mectizan, for human use to countries mainly in Africa which would otherwise not be able to afford it to control onchocerciasis, a related filarial disease. Glaxo-Smith-Kline later similarly pledged to donate albendazole for use in developing countries.

However at the time I was working in Malaysia in the early 1970s, ivermectin was not yet available and DEC was the only pharmaceutical drug certified for use against lymphatic filariasis. As would be expected in a drug developed and originally formulated for animal use, there were a lot of undesirable side effects in patients. One of the biggest challenges in the pharmacological treatment of lymphatic filariasis is the difficulty of drugs to reach the adult worms in their relatively protected location in the lymphatic system. Extremely high dosages of the drug would be required to reach and eventually accumulate in the slow flowing lymph to kill the adult worms and the toxicity of DEC on the rest of body would not be tolerated by the patient.

Therefore the strategy to eliminate filariasis was aimed not so much at treatment to kill the adult worms in infected persons but rather to break the cycle of transmission so that new individuals were not infected. This would essentially involve clearing the blood of microfilariae by giving everybody living in an endemic area a dose not high enough to cause severe side effects, yet sufficiently high enough to clear microfilariae from the blood circulation. This strategy is called mass chemotherapy. The idea was that even if the drug dosage did not completely eliminate all the circulating microfilariae in the entire population, if it decreased the availability of blood microfilariae to be picked up by mosquitoes to a critical level, there would eventually not be enough infectious mosquitos to maintain the cycle in that particular community. Thus eventually the disease would disappear as no new cases would occur and the old cases would fade away naturally over time due to attrition.

At the time I started working with Mak Joon Wah in the filariasis control programs, he was experimenting with different drug regimes in the mass chemotherapy campaigns to control lymphatic filariasis in Malaysia. He was attempting to compare how effective a 12 day dose of DEC, as recommended then by the World Health Organization (WHO), would clear microfilariae and whether a single high dose could achieve the same results. In low resource settings, obviously it was much more difficult and costly to attempt to manage treatment programs which had to ensure compliance in dispersed rural communities over a 12 day period. Single dose regimes would be much more cost effective. Furthermore DEC had unpleasant side effects ranging from nausea to fever and skin rashes.

This was my introduction to an actual boots-on-the-ground public health intervention and I found myself returning to Pondok Tanjung, in the district just outside Taiping where I was born. These programs, while very serious in intent, exuded an atmosphere not unlike a travelling carnival. We would descend upon a relatively isolated and quiet rural community, where rarely anything exciting happened. We would arrive with our vans and cars loaded with microscopes, dissecting equipment, medical supplies, and field technicians busy unloading and setting up the tables with lamps and microscopes to conduct blood examinations, dissecting scopes on benches to examine mosquitos, all kinds of bottles filled with reagents, boxes full of syringes, specimen vials, packets of gauze, bandaids, wipes etc. Since blood examinations were done late at night when microfilarial counts were highest, we sometimes took over the village center or local school canteen, and when the staging area was lit up at night it must have looked like the circus had come to town! To add to the general festive mayhem, usually Mak's teams would set up a projector and show educational movies about the filarial parasite, clinical signs of disease, parasite life cycle and the role of mosquito vectors and these would attract the attention of the village children. Sometimes short cartoon reels were interspersed among the public health movies to keep the children enthusiastic. As the children were the main agents of change in their communities, they would be encouraged to view their own blood slides under the microscope to look for microfilariae and this was an important aspect of community education to encourage active participation. Otherwise the perennial blood examinations and taking of medications with unpleasant side effects would soon have taken its toll and we would have been made very unwelcome. It was amazing that the children would line up giggling and laughing to have their blood drawn with finger pricks which must not otherwise have been a very pleasant experience (Fig. 6.9). I soon learned how a well-planned public health intervention, organized under carefully thought out social and behavioral considerations, could be efficient and effective.

Eventually other drugs such as albendazole and ivermectin became available for human use in addition to DEC, and they were instrumental in contributing to greater

Fig. 6.9   Filariasis blood survey, Pondok Tanjung, Malaysia circa 1975. Picture by author

success in the control of lymphatic filariasis. Studies in laboratories, field centers and community programs in Africa, Asia and Latin America similar to Mak's concluded that a single dose of two drugs administered together, albendazole with either ivermectin or DEC, can be 99% effective in eliminating microfilariae for a full year after treatment. This drug strategy was promoted globally as the Mass Drug Administration (MDA) by the WHO and remains the standard adopted by all countries.

Social and cultural stigmas, associated with a disfiguring condition such as elephantiasis, also complicated programs to control lymphatic filariasis. An incident that occurred in one of the smaller rubber plantations outside Pondok Tanjung reminded me of this to this day. When we surveyed the rates of people with microfilaria, looked at the prevalence of infections in the monkeys and calculated the proportion of infectious mosquitoes in that community, we were puzzled that we had not observed anyone with elephantiasis while we were there. According to our estimates, we should have expected to observe at least one or two cases of elephantiasis in that sample population. So we enquired with the *mandor* (foreman) of the rubber tappers in that plantation. He said he had not seen anyone with *untut* which was the local name of elephantiasis, but he would ask one of the retired tappers who still did odd jobs under his supervision around the estate whether he had heard of anyone around who might have *untut*. "Ya, ada" he answered, "Yes, there is". He then pulled up his sarong a couple of inches from his ankles and displayed early signs of elephantiasis which he had kept well hidden during the last few years while he was working there (Fig. 6.10). He apparently did not want to be known as

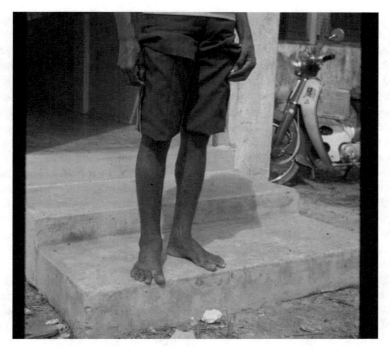

Fig. 6.10 Early stage of lymphatic pathology on right leg of rubber plantation worker. From collection of Mak Joon Wah

someone who had filariasis as he was afraid of being laid off. The fact that the *mandor*, who had been interacting with a person with elephantiasis on a weekly basis and yet had not known about it, demonstrated the power that social stigma had in confounding public health programs. It must also have been why I had not encountered elephantiasis in all the years when I was growing up in Taiping nearby.

Alas despite the well designed and scrupulously planned programs instituted by Mak and his colleagues, sub-periodic *B. malayi* in Peninsular Malaysia proved to be very intractable due partly to the persistence of transmission among the *Presbytis* monkeys. The mosquitoes maintained continual reinfections of people living in the areas by transmitting the parasites from the monkeys to humans, even after successful mass chemotherapy had eliminated the disease in the human population. Eventually it was economic development, and the socio-occupational changes brought about by urbanization, that probably led to the demise of filariasis by sub-periodic *B. malayi*. As the younger generation moved away from work in the agricultural sector to the towns to work in factories and air conditioned offices, occupational exposure to infectious mosquitos was drastically reduced. The cycle of transmission was eventually broken in many places because there were few new individuals working under the rubber trees, and exposed to mosquito bites, becoming infected. Developmental economists preach about an inverse correlation between the number of kilometers of new highway construction to the number of cases of infectious disease in developing countries, and it seems they are right.

A different strain of Brugian filariasis in Malaysia at that time was caused by periodic *Brugia malayi*. This strain was mainly found in the areas of rice cultivation in the coastal flatlands in northwestern Peninsular Malaysia in states like Kedah, Perlis and Penang. The mosquitos that vector this strain were mainly *Anopheles* species such as *Anopheles campestris* which breed in the open expanse of flooded rice fields, a totally different habitat from the riverine swamps of the sub-periodic strain in Perak. These mosquito larvae hang from the water surface, thus making them more vulnerable to conventional pesticide control measures. Furthermore, *Anopheles campestris*, the main vector mosquito species of periodic *Brugia malayi* happened to be also an important vector for malaria. So the better funded malaria control programs had been doing our job for us by eliminating the vector for periodic filariasis, and this strain never did pose the intractable problem like its sub-periodic cousin. This strain also does not have an animal reservoir host, thus making it easier to ensure that reinfection did not occur once the disease had been eliminated in the human population.

Although *Brugia malayi* is the important filarial parasite in Malaysia and other parts of SE Asia, such as Indonesia, Philippines, Vietnam, Thailand and southern India, *Wuchereria bancrofti* is the most important species among the lymphatic filarial parasites worldwide. The WHO estimates that as many as 947 milllion people in 54 countries are infected with lymphatic filariasis, at least 90% of which are due to *W. bancrofti*. This species extends its global footprint across Africa, Latin America and Asia and much of it occurs in the densely populated urban areas, among the sprawling city slums in developing countries. Movements of millions of migrants who relocate from rural areas into the major cities to seek employment, overwhelm the resources of city planners rapidly and the lack of water supply, sewage and drainage meant that habitats for the vector mosquitoes such as *Culex quinquifasciatus* become semi-permanent features of the mega squatter settlements. *Culex quinquifasciatus* mosquitoes are adapted to breed in the slums among the clogged drains, polluted canals, pools of dirty water, and the runoffs from rudimentary latrines which are common in regions where tropical rainstorms occur frequently. In many instances the mosquito larvae could be found breeding in stagnant pools literally within a few feet from the makeshift squatter huts and it is obvious that it would take wholesale urban redevelopment with proper housing, sanitation and water supply to eradicate Bancroftian filariasis, and mere vector control and chemotherapeutic MDA by themselves would be insufficient to control the disease.

In contrast to the massive urban challenges of Bancroftian filariasis worldwide, in Malaysia the strain of *W. bancrofti* is found mainly in sparsely populated remote rural areas near to the jungle in Sabah and Sarawak. Those are vectored by Anophelene mosquitoes and are relatively a minor problem unlike the urban disease found in other countries. Another species of lymphatic filarial parasite is *Brugia timori* but it only occurs in a few provinces in Indonesia and does not have the importance or geographical distribution of the other two species.

Chapter 7
River Blindness: And It Was About Time

How I ended up living in the United States was related to my work on filariasis. It appears that my peregrinations around the world had all been somehow dictated by many of the parasites I had encountered! I first met Willy Piessens, who was a Harvard professor working on *Brugia malayi,* in Chicago at the International Congress of Parasitology in 1978 and we soon became good friends. Willy had invited me to Berghoff, an iconic Chicago establishment, for dinner and we talked about filarial disease over *sauerbraten* and tankards of *hefeweizen.* During the late 1970s and early 1980s, Willy and his wife Pat, affectionately known as P, had been spending their summers doing field research in Indonesia on *Brugia malayi* and on a couple of occasions they visited us at our home in KL. When I mentioned on one of their visits that I was due for a sabbatical, Willy invited me to spend it at his laboratory at the Harvard School of Public Health in Boston. So in the winter of 1984, funded by a WHO grant to work on filariasis, I bundled up my family and we flew into Boston. The generosity and hospitality of Willy and P knew no bounds as they put up my wife Lucy and I, and our two young daughters to stay in their beautiful home filled with fragile antique Javanese batik hanging on the walls and shelves of precious and exquisite Ming porcelain which they had collected while living in Indonesia and in Guizhou Province in China. Our girls Shiamin and Shialing had charmed their way into the hearts of P and Willy, and P soon found an apartment for us in Brookline and even pulled strings to somehow have our girls enrolled at the Edward Devotion School even though school term had already started when we moved there. I was to later discover that President John F. Kennedy had attended that same public school and to this day I am totally astounded by how welcoming and generous the United States is to its immigrants. Our daughters were later to graduate with doctorates from Harvard and Yale on full scholarships and I still cannot imagine any other country that would truly open its doors that wide to people from other countries.

Willy had studied human populations in *Brugia malayi* endemic areas of Indonesia to understand human immune responses to microfilariae by serum-dependent cellular immune reactions as well as those mediated by drugs like DEC. He had also

© Springer International Publishing AG, part of Springer Nature 2017   75
B. H. Kwa, *The Parasite Chronicles*, https://doi.org/10.1007/978-3-319-74923-5_7

discovered mechanisms of specific immunodepression in microfilaria-positive individuals in endemic human populations and was among the first to report the existence of serum suppressor factors in microfilaria-positive individuals which appeared to interrupt the host effector mechanisms to the developing parasites. These studies ultimately opened the whole area to our understanding of the immunopathology of human lymphatic filariasis, and its complexity as a disease that has such a broad spectrum of clinical presentations.

During my sabbatical in his lab, Willy had introduced me to Ann Vickery DeBaldo who was a professor at the College of Public Health in the University of South Florida in Tampa. Ann had pioneered a unique athymic mouse model which allowed her to dissect the various mechanisms of the immune system in a living host infected with *Brugia malayi*. Because these mice are genetically born without a thymus, they do not have functional T cell mediated cellular immunity, and are essentially immunocompromised like individuals with AIDS. These mice happened to be also hairless, and are known as "nude mice" (Fig. 7.1). By selectively reconstituting various cellular sub-types which have been either exposed to specific parasite antigens in vitro or by actual infection in vivo, Ann was able to reconstruct the specific roles of the various components of the immune response to lymphatic filarial disease. This was an elegant technical breakthrough that substantially contributed to knowledge of *Brugia malayi* infections, and I was privileged to be able to participate in research with this model.

However, I had to return to Malaysia at the end of that year to fulfill my contractual obligations at the university to which I was tied because of the sabbatical I had taken. Two years later in 1986 after my contract had expired, I finally decided

Fig. 7.1 Brugia malayi infected nude mouse with dilated lymphatics. From collection of Ann Vickery DeBaldo

to emigrate to Tampa, Florida to take up a position at the USF College of Public Health and to work with Ann on her research on *Brugia malayi*. I was to remain at USF for the rest of my academic tenure as a professor.

Those were wonderful years, when I embarked on a new learning curve in understanding the ideas and vocabulary of public health as practiced in a modern post-industrial society such as the United States, with formal disciplines in health policy and management, health economics, occupational medicine and ergonomics, indoor air monitoring, industrial hygiene, etc., most of which were new and bewildering to me. I had been more familiar with the other core areas of epidemiology, biostatistics, environmental health and the behavioral sciences, but had to still educate myself to function in a new environment. It helped that the many friends and colleagues that I had gotten to know at USF had been welcoming and generous in making my transition to a new country so painless.

I continued to enjoy teaching and because the USF College of Public Health was then an all graduate school, it was a totally new experience for me. Many of the students in my classes were public health professionals who worked in the Florida Department of Health, others were physicians with years of clinical practice behind them, some were nurses with field experience during stints with *Médicins Sans Frontières* in places like Afghanistan, still others were returned Peace Corps volunteers who had worked with community health workers in places like Peru, Uganda or Nepal. I had to make my classes in parasitology and tropical diseases interesting and entertaining enough for them, as in many cases their experience and expertise made them a tough crowd to please! I must have succeeded despite the odds as the students voted to award me the Distinguished Teaching Award soon after I had started teaching. In some ways this book was a result of many of my students urging me to write down the stories I told in my classes about the parasites and parasitic diseases.

While most Americans would consider filarial parasites as exotic infectious agents that they would never encounter in the US, these parasites are in fact closer to home than they think. One day my doctoral student, Mike Pentella, who was then a senior clinical microbiologist at the Lakeland Regional Medical Center brought in a specimen that an ophthalmologist David D'Heurle had removed from a patient in Lakeland. In an eye examination to treat a minor inflammation, the doctor had seen something moving and had proceeded to surgically remove an inch long worm from the sub conjunctival tissue of the patient. The patient had never traveled out of the US but had reported to being routinely exposed to mosquitoes where she lived. Upon examination, I identified the worm as *Dirofilaria tenuis* which is a mosquito transmitted filarial worm commonly found among raccoons in Florida. Most dog owners in the US are familiar with *Dirofilaria immitis*, the dog heartworm, which is closely related to this parasite. Thus although the major human lymphatic filarial diseases are not found in the US, apart from occasional cases in immigrants who acquired them from their home countries, infections by indigenous zoonotic filarial parasites such as *Dirofilaria tenuis* do occur sporadically in our backyard in a subtropical state such as Florida.

From my vantage point in Florida while working at USF, the entire continent of South America beckoned below from just beyond the Caribbean, and it was there that I first encountered the disease known as River Blindness.

* * *

My plane had been delayed in Caracas and by the time it landed in Puerto Ayacucho, we were more than 2 h late. It was six thirty, late in the afternoon in November 1989, and dusk was beginning to set in. In those days the airport in Ayacucho was just a small wooden building with a corrugated tin roof and the single airstrip was surrounded by jungle. Most of the other passengers were met by family and friends and they soon clambered with their luggage onto cars and the flatbeds of small pickup trucks and disappeared down the dusty track that led out of the airport. I was expecting someone from CAICET (which is the acronym in Spanish for Venezuela's Amazonian Centre for Research and Control of Tropical Diseases) to pick me up and had turned down the two cab drivers trying to steer me into their cabs. I had just been ripped off for a large sum by an unscrupulous cabbie the previous day when he had driven me from the Caracas International Airport by the coast, all the way up to the mountains where the city of Caracas was located. I had only wanted to have a lay-over at a hotel near the airport to catch the connecting flight to Ayacucho the next day. Therefore I had become very suspicious of being maneuvered and manipulated by another cabbie once again. Soon however it became obvious that there would not be anyone coming to pick me up, as all the other passengers had long since departed and the airport was totally deserted. The lone airport worker started to lock up the little storefront of the airport kiosk selling cigarettes and candy, as he seemed to be both the airport official as well as the store manager, and he was beginning to turn off all the lights. He looked at me quizzically and asked something in Spanish which I could not understand but I heard the word "taxi" and so I quickly nodded "Si, si, taxi!" He rolled his eyes and pulled a face at this El Chino Loco who had just waved away two taxis earlier but reluctantly unlocked his store and shouted into a phone, his voice echoing too loudly in the already silent and desolate building. He then relocked his store with a clang, turned off the lights, gave me an awkward smile, and headed to his motorbike. With one final pitying look in my direction, he kick-started his bike, which exploded to life in the deafening silence, then turned on the headlight and puttered off. As the sounds and light of the motorbike disappeared into the gloom, the incessant buzzing and shrieking of the surrounding jungle suddenly grew very loud. It was by now almost totally dark, with just a few remaining streaks of purple and vermilion glowing in the horizon above the black silhouette of the forest.

I stood there with my single duffle bag in the dark outside the airport, and the mosquitoes had started to bite, and felt the complete fool. I had been invited by a colleague and friend Izaskun Petralanda who had done her post-doctoral research on onchocerciasis in Willy Piessens' lab at the same time when I was there. She had organized a workshop at CAICET where she worked and had invited me to speak on *in vitro* culture methods for filariasis, and I had flown in from Tampa. Willy was going to be there as well, so I had looked forward to the trip as I had not been to Latin

America before. Puerto Ayacucho was in the State of Amazonas in Venezuela, right beside the mighty Orinoco, at the border with Colombia just across the river. Its location deep within the Amazonian Basin made it an ideal place to study tropical diseases especially onchocerciasis, as the disease was endemic among the Yanomami tribes living in the densely forested region there.

Onchocerciasis, also known as River Blindness, is caused by a filarial parasite *Onchocerca volvulus*, related to the lymphatic filarial parasites we had discussed in the previous chapter. Unlike the lymphatic filarial parasites however, the adult worms of *O. volvulus* do not live in the lymphatics but in the subcutaneous tissue under human skin (Figs. 7.2, 7.3, 7.4). The disease is spread across many parts of sub Saharan Africa, but was brought over into South and Central America with the slave trade. Worldwide the WHO estimates that 18 million people are infected and 120 million live in endemic areas. As many as 300,000 people are blinded by onchocerciasis and another 800,000 people have visual impairment due to the disease and 6.5 million suffer from severe itching brought about from the parasite in their skin.

The insect vector that carries them are blackflies belonging to the genus *Simulium* and these are the irritating tiny flies that swarm around you and give you an extremely itchy and painful bite. They are the ruin of many a summer's day for canoeists and fly fishermen in the northern US and Canada who experience their peculiar torment, although fortunately they do not harbor the parasite in those places.

In many parts of Africa however the blackfly species *Simulium damnosum* and others in that species complex are the main vectors for the parasite *Onchocerca*

Fig. 7.2 Onchocerca volvulus (Source: Encyclopedia of Parasitology Ed. Melhorn 2016, Fig. 2)

Fig. 7.3 Onchocerca
volvulus (Source:
Encyclopedia of
Parasitology Ed. Melhorn
2016, Fig. 3)

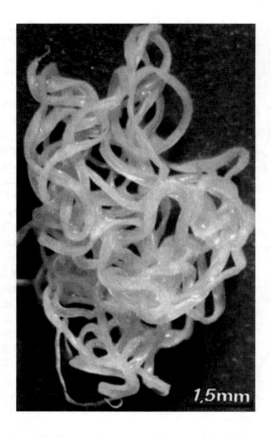

volvulus, and in South America, such as in the region around Puerto Ayacucho,
different species such as *S, guianense* and *S. exiguum* are the predominant vectors. In
Central America flies belonging to yet another species, *S. ochraceum,* are predominant. The blackflies breed in fast moving streams and riverine rapids, particularly
where the water rushes over boulders, rocks and vegetation to create white water
turbulence. The fly larvae are filter feeders and they require a constant stream of
highly oxygenated water to thrive, as well as aquatic micro-organisms for food.

In certain places in Africa where vectors like *Simulium damnosum* tend to breed
in relatively large rivers, vector control using insecticides such as Temephos was
originally quite successful. However the land area covered was very large and
weekly aerial spraying, that had to be sustained over a period of 14 years to totally
eliminate transmission, proved very challenging. Furthermore they soon ran into
problems when it was discovered that the blackflies migrate over extremely long
distances and that they soon also developed resistance to the insecticide, rendering
vector control for control of onchocerciasis not feasible in many parts of the world.
In Central and South America, the disease is found in widely distributed and
relatively sparsely populated areas, where the vectors can breed in isolated small
streams and rivers, usually covered under a canopy of trees in thick jungle. This

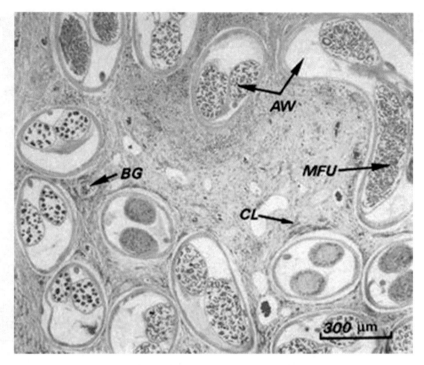

Fig. 7.4   Onchocerca volvulus (Source: Encyclopedia of Parasitology Ed. Melhorn 2016, Fig. 5)

presents yet another set of challenges for vector control as aerial spraying was not even possible.

Adult blackflies feed on mammalian blood, but when they obtain blood from our skin, the blackflies do not use a proboscis to puncture the skin and draw up the blood like mosquitoes do. Instead they lacerate the tissues of the skin using their mouth-parts like a food blender and then lap up the blood gathering in a pool on the skin surface. That is why in textbooks blackflies are referred to as "pool feeders", and that is why their bites are especially painful. The infective larvae will then migrate from the fly's mouth into the wound and penetrate into the subcutaneous layer below and settle in. There the parasite larvae will develop and undergo the molts of their cuticle like their filarial cousins and eventually mature into adult worms. Host inflammatory defenses surround the adult worms with a thick capsule of fibrous tissue and the adult worms will live within these nodules, also called onchocercomas, for the duration of their lives in human skin. Often several worms will be encapsulated within the same nodule, and if there are mature male and female worms together, they will mate and produce between 1000 and 3000 microfilariae daily. The microfilariae secrete powerful tissue-dissolving enzymes, such as collagenases and proteases, which would allow them to penetrate through the capsule wall and out of the nodule and migrate in the skin (Fig. 7.5).

Fig. 7.5 Onchocerca
volvulus (Source:
Encyclopedia of
Parasitology Ed. Melhorn
2016, Fig. 4)

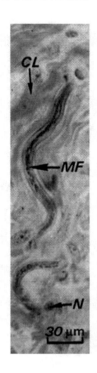

The gold standard for diagnosis of onchocerciasis depends on skin snips, whereby a small volume of skin, usually 1–2 mg, will be removed either by scalpel or a specialized skin biopsy punch. The skin sample is incubated in saline to determine if there are microfilariae present which would migrate out into the incubating solution. More sophisticated techniques such as PCR methods that look for the parasite DNA exist but are often not readily available in remote low resource settings.

Sadly the migratory activities of microfilaria often include invasion of the eyes, and here the microfilariae can damage the cornea, retina and sometimes the optic nerve leading to vision loss. When a blackfly next feeds on the skin, the microfilariae in subcutaneous tissue will migrate into the blood pool and be sucked up into the midgut of the fly. The microfilariae will then penetrate the gut wall and migrate into the thoracic muscles of the fly where they will develop, after two molts, into infective third stage larvae. These will migrate then to the mouth parts of the blackfly and lay there waiting for the next feeding of human blood to resume the cycle.

Just when I had started to despair, as I stood waiting alone in the dark by the Ayacucho airstrip, a pair of dim headlights appeared in the distance. As it drew up I noticed that it was a ramshackle rust bucket of an off-license taxi and the driver was an unshaven Mestizo of around forty, with wild looking eyes and wilder looking hair. A shirtless young boy, who was in the front passenger seat, bounded out and untied the piece of rope that was holding the back door of the taxi in place. He pulled the door open and smiled, waved me in, took my bag to put in the trunk, and slammed the lid a couple of times to close it. As I sat in the back, he came back and retied my door with the rope from the outside and hopped back into the front seat as the taxi started to rattle down the track. I said the only name I knew to instruct the driver about where I wanted to go, "Caicet. . . . . .CAICET!" I shouted. I hoped I pronounced it right. He turned to the boy, mumbled something and they both laughed. I had no idea if what I said meant anything to the driver, but he eyed me in the mirror and nodded and laughed. He also smelled like he might have had a few drinks already, but by that time I could not be sure of anything. After leaving the airstrip, the dirt track turned into the jungle and by then night had descended and it was totally dark in the forest. From the weak light of the taxi's headlights, the shadowy branches of the enclosing trees gave the impression that we were moving through some sort of ghostly tunnel. As our car approached, the deafening sounds of the screeching jungle would cease temporarily while we rumbled through, but would immediately resume with a vengeance behind us as we passed. I could just sit and wonder where on earth we were going.

Just when I had started to wish that I had been back with the Caracas taxi that had ripped me off, the driver stopped the car right in the middle of nowhere, turned to the boy and said something which I could not understand. The boy opened the door and went to the back and took something from the trunk. Is he taking. . . .my bag? Oh. . .no, it's worse! He re-appeared holding up a very sinister looking machete in one hand and was holding something else in the other hand which I could not see. Oh . . . ..God, so this was it. This was how it would all end! Before I could gather my thoughts, the driver abruptly turned around and looked me in the eye and asked "Que hora es?" Now, I must confess that at that time my Spanish was totally non-existent. He tried again, "Tiempo. . . . tiempo?" and pointed to my watch. Ah. . . .that's it. . . .whew!. . . . he only wanted my watch after all! So I took my watch off and handed it to him, which somewhat startled him. He took one look and then returned my watch to me. It was not what I had expected, and we were now both equally mystified! I noticed that the boy had disappeared by then and the two of us just sat there in the car in awkward silence, surrounded by the cacophony of the jungle. Minutes passed. So what was going to happen next? Will the boy be back soon with a whole village of men with torches, ropes and machetes? What?

The driver heard a sound in the bush and quickly opened the door, went to the front of the car and cranked open the hood. The boy was back and this time I could see that he was holding a bucket full of water and handing it to the driver. The radiator had apparently overheated and the boy had been sent to cut his way through the under-growth with the machete, to some jungle stream to fetch some water. Oops . . . .. that was embarrassing!

Soon we were off again, the bucket and machete safely stowed away in the trunk. The driver told some long story to the boy and they both roared with laughter, with the boy shyly making sidelong glances back at me and then doubling up and laughing hysterically again and again. Then we turned a corner and there was the magic word C.A.I.C.E.T suddenly appearing on a low wall as our headlights swept across it. I joined in the general laughter in cathartic relief, slapping the driver on his shoulder gratefully. Izaskun and Willy who welcomed us were bewildered to find all three of us in tears, breathless and still trying to stop laughing, and wanted to know what could have happened in that taxi that would have been that funny. They must have also wondered why I had tipped that taxi driver so extravagantly, but all I said was that the generous tip was for the very funny Yanomami jokes the driver told in the journey from the airport. This was my introduction to Latin America, a region that I have grown to love over the years since.

The research that Izaskun was doing at CAICET at that time mainly involved the immunopathology caused by *O. volvulus,* such as the role of collagenase enzymes secreted by their microfilarie in damaging skin and ocular tissues. However CAICET was also an important center for testing the efficacy of pharmaceutical drugs to treat onchocerciasis. The situation with onchocercal disease is quite different from lymphatic filariasis and is far more challenging. In the case of lymphatic filarial parasites, the main causes of pathology were the adult worms in the lymphatics, not the microfilariae. However in onchocerciasis, the adult worms that reside in the skin nodules do not cause any serious disease apart from the disfiguring nodular skin condition. In Guatemala, they trained squads of paramedical workers to go from village to village to perform nodulectomies. They would make incisions in the skin, usually in the scalp, where the nodules were and carefully removed the inner capsules with the adult worm without puncturing them. Otherwise the sudden release of copious amounts of parasite antigens would cause anaphylactic shock which could endanger the lives of the patients. Before the advent of successful drug treatments this was one way to remove the adult worms and reduce the production of new microfilariae.

In onchocerciasis, it is the microfilaria stage that causes the most serious disease: intense itching dermatitis, making the skin rough and leathery as well as causing unsightly hyperpigmentation or depigmentation (called "leopard skin"). Atrophy of skin tissue also contributes to loss of skin elasticity which results in a condition known as "hanging groin". All these radical disfigurements will result in social stigma for the unfortunate infected individuals. However, microfilariae also invade the eyes and this is of the greatest concern, as persons living in endemic regions could start to experience vision loss as early as in their teens. Dying microfilariae in the cornea would cause toxic reactions which begin as punctate (pinpoint) opacities known as "snowflake" keratitis, but repeated deaths of more migrating microfilariae would start irreversible lesions in the cornea known as sclerotizing keratitis. Microfilarial damage is not limited to just the cornea as further damage in the anterior chamber of the eye would result in cataract and glaucoma, and pathology in the posterior chamber will manifest itself in atrophy of the retina as well as the optic nerve. Thus this progressive vision loss would eventually result in complete blindness after a few years of exposure in endemic areas.

Whereas living microfilariae would cause some lesions, dead and dying microfilariae will provoke even more serious pathology due to host inflammatory reactions to the antigens released by dying microfilariae. So the one thing you would want to avoid most is to cause killing of microfilariae in ocular tissue by a drug such as DEC. It is therefore crucial to find an alternative drug that would circumvent this treatment dilemma. Ideally we need to come up with a drug that eliminates the adult worm, yet does not precipitate microfilarial death. Fortunately it was discovered that the adult worms of *O. volvulus* depend on a symbiotic bacterium called *Wolbachia* that lives intracellularly within the worms and which are essential to their survival. Thus parasites have their own parasites!

The *Wolbachia* bacteria are found abundantly in the parasite's female reproductive system, and appear to be transmitted into the developing embryos that become microfilariae, and these bacteria play an important role in the survival and reproduction of the parasites. The opacities in the cornea as well as the other lesions inside the eye chambers are a result of these host-parasite immunopathological cellular reactions. This was an important discovery as we can now use antibiotics such as doxycycline to destroy the *Wolbachia* which would in turn inhibit adult worm development, block embryogenesis and reduce fertility, and thus diminish the production and survival of microfilariae. The sterile adult worms will slowly die and disintegrate within the nodules eventually without causing any more disease. Currently the recommended dosage in the MDA regimes indicated for onchocerciasis is 200 mg of doxycycline daily for 6 weeks, which sterilized the adult female worms as well as reduced the number of microfilariae by curbing their production. There have been debates about the practicality of successful implementation of a pharmaceutical regime in the field that requires 6 weeks of compliance by the entire population at risk. However, the terrible suffering caused by onchocerciasis is apparently enough to result in 97% compliance in a population of 13,000 people in trials in Cameroon. Nevertheless, the search for an optimal drug continues as doxycycline is contraindicated for children and pregnant women, and therefore trials of other pharmaceutical agents such as rifampin are still on-going by the time this book was written.

Much of the work that the CAICET team does today focuses on the design of control programs that take into account the complex social structure of the semi-nomadic Yanomami and their spatial interaction with the Amazonian jungle as well as social contact with other tribal communities (Fig. 7.6). These social interactions

Fig. 7.6 Author in front of Yanomami dwelling, Puerto Ayacucho, Venezuela 1989. From collection of author

form an interconnected network of nodes in the jungle between different tribes which make it extremely difficult to prevent reinfections of previously treated individuals by untreated populations as they migrate along the little known labyrinth of forest tracks. This heroic work by researchers at centers like CAICET, working among the indigenous peoples of the Amazonian region, is often unheralded and little known outside the small circle of parasitologists, but it may make a big difference in the lives of these marginalized communities who are still seriously affected by River Blindness.

* * *

Another one of the many problems with pharmaceutical treatment for ochocerciasis and lymphatic filariasis in parts of West and Central Africa is because of another related parasite called *Loa loa* which causes the infection known as loiasis. The parasite lives in human subcutaneous tissue but instead of being encapsulated within nodules like those of *Onchocerca*, they migrate freely under the skin, and sometimes when they migrate near the surface, may actually be seen moving just beneath the skin. Frequently they would migrate in the subconjunctival region of the eye and across the field of vision of the infected individual (Fig. 7.7). British colonial plantation owners must have thought that they had spent too much time in the midday sun in Africa, and had gone stark raving mad when huge serpents crossed their very eyes while they were enjoying their gin tonics and other sun-downers on the verandah. Although highly disconcerting and sometimes accompanied by transient inflammation, there is usually no permanent damage to the eye—this is yet another difference in pathology from the disease caused by *Onchocerca* infections. A clinical condition known as Calabar swelling may also present itself usually in the upper and lower limbs, and this is a result of allergic reactions to the metabolic products excreted by the migrating worms. The most serious problem however is the high microfilarial loads that can occur with loiasis, and the fact that microfilariae of *Loa loa* can be densely found spread among various tissues including peripheral blood but also in spinal fluid, sputum, urine, and in the lungs. Thus there is a high risk that when individuals, living in areas with co-infections of loaisis and other filarial infections, are treated with DEC or ivermectin, as it would precipitate the sudden mass killing of *Loa loa* microfilariae and this can result in fatal encephalopathy. Although the cause of encephalopathy in loiasis is still poorly

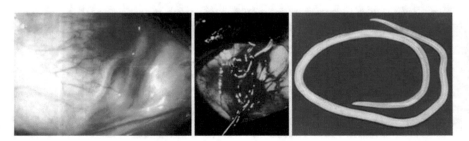

Fig. 7.7   Loa loa (Source: Encyclopedia of Parasitology Ed. Melhorn 2016, Fig. 1)

understood, it appears that because of the high density of microfilariae in the tissues, their sudden death could cause an immediate increase in permeability of the capillary walls of blood vessels in the brain. This appears to result in the drugs and antigen-host protein complexes crossing the blood-brain barrier and causing neurotoxicity. Thus the presence of encephalopathy caused by loiasis complicates the control strategy of MDA in such areas where *Loa loa* coexists with the other filarial parasites.

The vector for *Loa loa* is the deerfly belonging to the genus *Chrysops* which presents another problem that confronts vector control strategies. *Chrysops* breed in isolated and remote breeding sites in the forests and they are frequently spread thinly over very large areas. The eggs are laid on decaying vegetation on the soil surface and the larvae that emerge will bury themselves into the mud or damp soil making them inaccessible to conventional insecticide spraying. These deerflies are also pool feeders, like the blackflies, but because they are much larger, their bites are very painful. Chrysops are also day biters and therefore the use of insecticide treated bednets to prevent loiasis is also ineffective. Therefore apart from using repellents there is presently no effective control for loiasis.

* * *

Other minor related filarial parasites include mansonellosis which is caused by several parasites in the genus *Mansonella*. Examples are *Mansonella streptocerca* which is found only in Africa; *Mansonella perstans* in both Africa and South America; and *Mansonella ozzardi* only in the Americas, from Mexico south to South America and in the Caribbean.

Mansonella streptocerca does not cause serious disease but symptoms may include skin manifestations such as pruritus, papular eruptions and pigmentation changes. Eosinophilia is commonly associated in all filarial infections and is therefore ubiquitous in mansonellosis. Infections by *Mansonella perstans*, while often asymptomatic, would present with symptoms such as angioedema, pruritus, fever, headaches, arthralgias, and neurologic manifestations. *Mansonella ozzardi* will usually be associated with symptoms that include arthralgias, headaches, fever, pulmonary symptoms, adenopathy, hepatomegaly, and pruritus.

Chapter 8
Intestinal Worms: Parasite Mysteries in Wintry Climes

In the summer of 1996, I happened to be visiting Washington DC. As I love visiting museums and art galleries, whenever I had time on my hands in Washington I would head to the National Mall and wander into the Smithsonian complex of museums which are of tremendous value, especially since they are free! My favorite is the East Wing of the National Gallery of Art, designed by I.M. Pei, with its sharp angles and dramatic geometric facades externally, and the bright glass and steel atrium within, looking out onto a cascade of water rippling over stone baffles on an inclined wall, like a waterfall. I often enjoyed the building itself as much as the exhibits within.

I had just spent the previous August in Beijing reading a paper at the 10th Anniversary Meeting of the Chinese Society of Parasitology in 1995. Whenever the sessions of that conference were over I would visit the numerous museums and galleries of the Chinese capital that were accessible via the Beijing metro. I found out that the majority of the local Chinese parasitologists at that meeting had the same idea, as most of them were from the provinces and had used the opportunity of the national conference as a perk to do some sightseeing. That was the first time I had visited Beijing, and that trip coincided with our whole family accompanying our daughter Shiamin to help her get settled into her dorms at the Beijing Normal University (known by the Chinese acronym Běishīdà). She had been selected from her Dartmouth College class to teach English at Běishīdà, and in turn she was to learn Chinese there. So besides reading my paper at the conference, I had plenty of time to visit museums and learn about the development and evolution of Chinese artistic styles from the early Shang period onwards to the late Qing dynasty.

So there I was in Washington DC, a year later, at the National Mall heading towards the National Gallery's East Wing where a huge banner announced an exhibition of "Olmec Art of Mexico". As I walked through the exhibits of Olmec sculptures, most of which were monumental pieces of stone and jade, the first thing that hit me was how eerily they reminded me of the Shang sculptures and jade pieces I saw a year earlier in Beijing! Many of the faces depicted in the Olmec sculptures of human figures, especially of children, were distinctly of East Asian physiognomy. But what was even more intriguing were the patterns on Olmec jade "celts" which

© Springer International Publishing AG, part of Springer Nature 2017
B. H. Kwa, *The Parasite Chronicles*, https://doi.org/10.1007/978-3-319-74923-5_8

were relatively small ornamental blade-like pieces which to my eyes looked very much like Shang artifacts, and some of the Olmec inscriptions had borne striking resemblance to Shang writing. This stayed on my mind long after my visit on that summer's day in 1996 at the National Art Gallery and it remained unresolved, like an itch that I cannot scratch even years later.

So what has it got to do with nematodes and hookworms? Well, one of the great debates in ethno-archaeology and anthropology has to do with how the Americas were populated by its indigenous peoples. The scientific consensus over the past several decades have been in agreement that the ancestors of Native Americans crossed over from Siberia across the land bridge that had connected the Asian land mass to Alaska. Beringia, as the land bridge is called, existed around 30,000 years ago, but was subsequently submerged a little over 10,000 years ago when sea levels rose and the Bering Strait was formed. There are alternative, less favored, theories that postulated trans-Pacific maritime migrations, with ancient mariners following ocean currents travelling from either mainland East Asia or from archipelagic Southeast Asia across the Pacific Ocean to South America and then diffused northwards. There is little disagreement today that there were probably several waves of migration from the Asian land mass and recent DNA evidence suggests that there are enough common genetic markers between peoples of the Altai region in Russia, that borders China, Mongolia and Kazakhstan, to Native American populations to make the Altaic connection difficult to dispute.

For parasitologists, the Beringia theory of migration from Siberia to Alaska and onwards to South America poses a conundrum. You see, hookworm eggs were found in human coprolites (fossilized fecal remains) from a mummified body in Minas Gerais, Brazil which was about 3000–4000 years old. Since the hookworm eggs predated the landing of Columbus by several millennia, if the cross-Beringia migration hypothesis is correct, the hookworms could only have come over with the original migrants, since hookworms like *Ancylostoma duodenale* were of Old World origin. Here lies the conundrum. Hookworms have obligate free-living larval stages that need to survive and develop in the soil for at least 5–10 days in warm moist conditions, which is why hookworms today are only found in the tropical or sub-tropical regions. So how could hookworms have survived the journey in the arctic and frigid conditions of the Beringia land bridge?

There are two species of human hookworms, *Ancylostoma duodenale* and *Necator americanus*. Under the microscope, hookworm eggs have thin shells and look delicate and fragile (Fig. 8.1). When they are passed out in stool the embryos within the shell are usually unsegmented or in the early stages of segmentation, in the two or four cell stage. They will require warm moist conditions to develop in the external environment for one or two days, when a fully formed larva will hatch from the egg shell. This first stage larva is called a rhabditiform larva and it will need to develop and grow, feed on bacteria, and molt twice, before becoming the filariform larva. This is the third stage infective larva which is now ready to infect the human host (Fig. 8.2). This entire process takes between 5 and 10 days, during which the larvae will actively move about on the soil surface, feed and grow, and they require warm moist conditions common in tropical or subtropical climates to survive during

Fig. 8.1 Hookworm
disease (Source:
Encyclopedia of
Parasitology Ed. Melhorn
2016, Fig. 2)

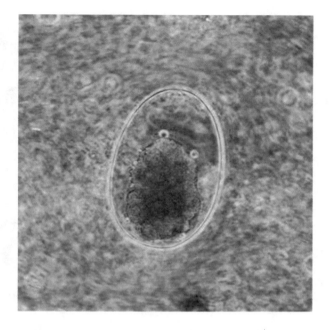

this stage. They are very susceptible to desiccation and cold weather and they will die quickly under those conditions. The infective larvae penetrate directly into the skin of bare feet, usually in the more delicate skin between the toes. Again it is usually in tropical climates where children and adults are likely to be walking barefoot and be exposed to soil contaminated by fecal matter.

After entering the skin, the larvae will find and enter into a small capillary or lymph vessel and from there enter into the blood stream. They will be carried to the heart and into the pulmonary circulation and eventually into the capillaries surrounding the small alveoli of the lungs. There they will leave the blood system and penetrate into the air sac and migrate up the small bronchioles, then to the trachea into the larger airways up the bronchial tree to the pharynx where they will be swallowed into the esophagus and then into the digestive tract. One of the passages of the book "Introduction to Parasitology" by Asa Chandler and Clark Read, which was my bible when I was studying parasitology at ANU, suggests that the chewing of betel nuts common in India and many other parts of South Asia may have been helpful in reducing the worm load of hookworm infection (page 429, ibid.). Betel chewers tend to hawk and spit out red blobs of phlegm all day long, and while it is an unhygienic and disgusting habit, they would constantly expectorate the hookworm larvae instead of swallowing them and allowing the hookworms to complete their life cycle. The authors further suggested that this may be the same reason why chewing tobacco was once considered a healthy habit in the U.S. South, since hookworms used to be prevalent there too!

During this period of their tissue migration, the hookworm larvae will molt two more times before they finally mature into adult worms in the small intestines. There the adult worms will attach themselves to the intestinal wall and feed on the blood,

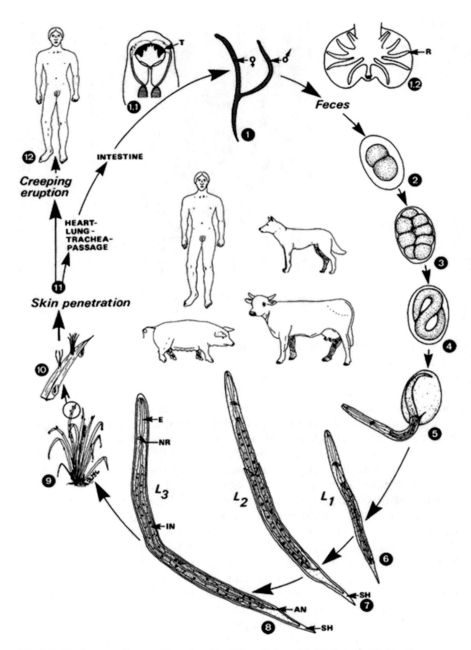

Fig. 8.2 Hookworms (Source: Encyclopedia of Parasitology Ed. Melhorn 2016, Fig. 1)

using specialized mouth parts both for attachment as well as for cutting into the blood vessels to draw blood (Fig. 8.3). The open mouth cavity of *Ancylostoma duodenale* looks like a snake's jaw with vicious looking fangs, while those of *Necator americanus* have broad curved blades like butcher's cleavers (Figs. 8.4 and 8.5). In some studies in humans infected with *N. americanus*, as much as 90 ml of blood loss per patient per day was estimated based on measurements of red cells labeled with Cr^{51} isotope. Hookworm disease is therefore associated with protein deficiency, anemia, iron deficiency and malnutrition, and although it does not cause any mortality, it is probably one of the most insidious silent public health challenges in the developing world as it retards growth and mental development among children (Fig. 8.6). Worldwide the CDC estimates that as many as 500–740 million people are infected. Treatment for hookworms include mebendazole, albendazole and pyrantal pamoate.

It remains unanswered why the larvae undergo such an elaborate tissue migration, since they essentially migrate via the blood into the lungs and then end up back into the intestines again where the adult worms live. It has been suggested that this tissue migration may be related to ancestral forms of the hookworm which had colonized different organ systems and required specific physiological stimuli to develop at different stages of its life cycle. It is very likely that nematode parasites had evolved from free living nematodes, of which there are many examples today, which had later adapted to a parasitic mode of life in a variety of hosts. There are examples of non-human hookworms of other mammalian hosts which accidentally enter human hosts, and lacking the correct physiological triggers will fail to develop normally and continually migrate in human tissues. For example, zoonotic infections from

Fig. 8.3 Hookworms (Source: Encyclopedia of Parasitology Ed. Melhorn 2016, Fig. 3)

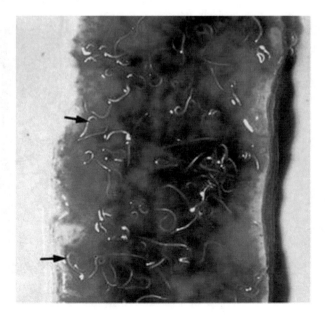

Fig. 8.4 Hookworms of
animals (Source:
Encyclopedia of
Parasitology Ed. Melhorn
2016, Fig. 2)

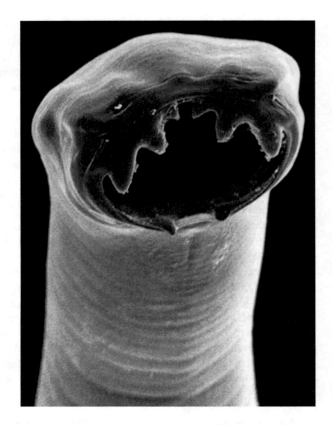

Fig. 8.5 Hookworms of
animals (Source:
Encyclopedia of
Parasitology Ed. Melhorn
2016, Fig. 3)

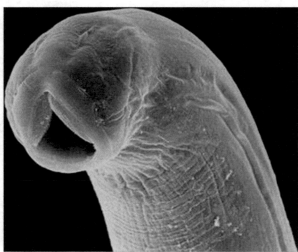

Fig. 8.6 Railway tracks running through village houses infested with hookworm disease, Sentul, Malaysia circa 1973. By author

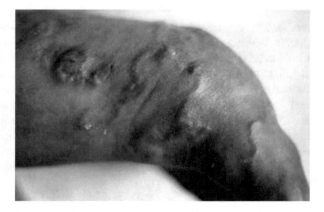

Fig. 8.7 Hookworms of animals (Source: Encyclopedia of Parasitology Ed. Melhorn 2016, Fig. 6)

hookworms of cats and dogs such as *Ancylostoma braziliense*, *A. caninum*, *A. ceylanicum*, and *Uncinaria stenocephala* cause cutaneous larva migrans when they accidentally infect humans. People are infected when they walk barefoot, sit on contaminated ground or lie on sand contaminated with animal droppings, and have the infective larvae penetrate their skin (Fig. 8.7). That is one of the reasons why dogs and are barred from public beaches; the other reason is because of toxocariasis which we will discuss later. Cutaneous larva migrans appears as red raised cutaneous tracks where the migrating larvae passed under the skin, and they cause itchiness and rash which may last several weeks, until the larvae eventually die.

Thus we return to the conundrum of how humans and hookworms could have first migrated to the Western Hemisphere. We have noted that the entire hookworm cycle is adapted to a tropical climate to which it has evolved, and would not have survived the wintry ice sheets that existed in the arctic climate of the Beringia land bridge. There remains the alternative theory of a maritime route across the Pacific Ocean that would have to be seriously considered because of the hookworm argument. Archae-ologists and Chinese experts on Shang dynasty artifacts have constructed a hypoth-esis that postulated that the fall of the Shang dynasty in China, which coincided with the later rise of the Olmec civilization in Mexico, could have precipitated migratory waves from the Asian mainland around that period about 3000–3500 years ago. Historical records about the chaos in the aftermath of the collapse of the Shang mentioned a spate of migrations across the ocean by remnants of the defeated Shang armies escaping down the Yellow River to the coast on China's eastern seaboard. That could account for the postulated linkage between the Shang and Olmec civilizations which I found so intriguing in the National Gallery of Art in the summer of 1996. Although highly speculative at this stage, if the hypothesis were ever to be substantiated by future new evidence, then parasitology and hookworm biology would have contributed no small part towards answering the decades-old question about the route the original ancestors of Native Americans had taken to populate the American continents.

∗∗∗

Another intestinal nematode that is a scourge for people living in the tropics is the roundworm *Ascaris lumbricoides*. It is estimated that there are 800 million–1.20 billion people infected worldwide, and like the hookworms because they do not directly cause death, tend to be under-reported in the mainstream media which prefer to emphasize more dramatic outbreaks of Zika or similar infectious disease *du jour* with every news cycle. However *Trichuris trichiura*, hookworms and *Ascaris lumbricoides* form what is referred to as the triad of intestinal worm infections in classic texts of tropical medicine, as together they account for a major disease burden in the tropics as measured by disability-adjusted life years (DALY) lost. The WHO estimates that the total prevalence of soil transmitted worm parasites exceeds 2 billion globally.

The eggs of *Ascaris lumbricoides* are totally unlike those of hookworms when they are examined under the microscope, as they have shells that are thick and rough with irregular bumps and a craggy look. The shell is actually as tough as they look, as this is the infective stage and the eggs are adapted to withstand the harsh and unforgiving conditions of the external environment for long periods (Fig. 8.8). When I prepared live specimens of *Ascaris* eggs for parasitology lab classes to study egg hatching for instance, I would actually store them in 1% hydrochloric acid and kept them alive for months in the refrigerator. They would also survive and continue to embryonate quite well in 10% formalin and 7% glacial acetic acid, which would normally kill everything else. Under normal climatic conditions, *A. lumbricoides* eggs would remain viable for years through all seasons in moist soil. In Chandler and Read's tome "Introduction to Parasitology" which I had mentioned earlier was our

Fig. 8.8 Ascaris, species of animals (Source: Encyclopedia of Parasitology Ed. Melhorn 2016, Fig. 2)

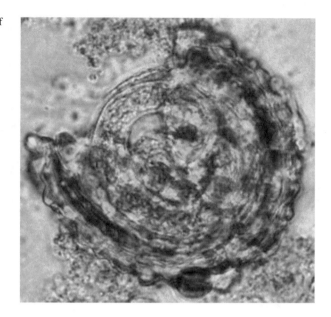

undergraduate textbook, there is a passage (page 453, ibid.) relating how an "enterprising German researcher seeded a plot of soil with *Ascaris* eggs; two persons ate unwashed strawberries raised on the plot each year for 6 years, and each year acquired a few *Ascaris*". That was obviously long before the era of informed consent and the institution of strict ethics review boards that regulate all forms of medical research today!

Ascaris lumbricoides eggs passed out from the human host require at least another 10–14 days to develop in soil under favorable conditions of temperature, moisture and aeration for the larva to develop and molt once within the shell. These mature eggs with the second stage larva then become fully infectious. When ingested by a human host they would hatch in the small intestines and penetrate mucous membranes of the intestines and enter the blood stream (Fig. 8.9). They then proceed to undergo the same tissue migration we had seen in hookworms, from the blood to the heart, then the lungs where they will burrow their way into the trachea, up the bronchial tree to the pharynx and be swallowed into the esophagus and back into the intestines. Eventually the adult *Ascaris* larvae would grow 10 times in size and become juvenile worms of about 3 mm. They will then mature and grow in the human small intestines into sexually mature female worms around 35 cm in length, and males of 30 cm. Each female worm can produce an astounding 200,000 eggs daily, thus ensuring that the likelihood of transmission of *Ascaris* under optimal conditions is extremely high.

During the pulmonary stage of larval migration, heavy infestations could potentially cause serious illness such as hemorrhages and even pneumonia, although there are seldom enough eggs ingested at the same time to cause serious pneumonia in

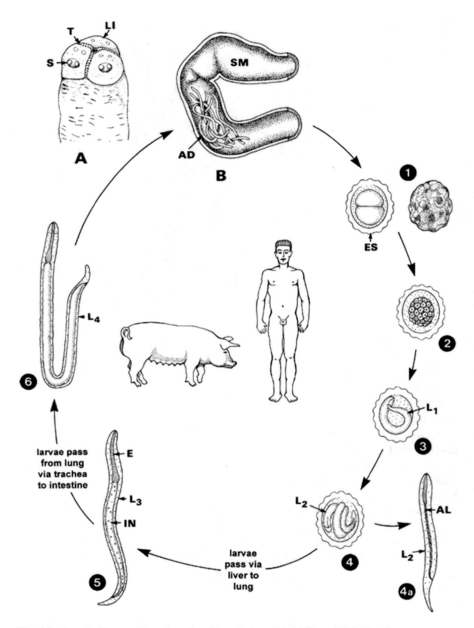

Fig. 8.9   Ascaris (Source: Encyclopedia of Parasitology Ed. Melhorn 2016, Fig. 1)

natural infections. However there was actually a spectacular court case in Canada in 1970 when an undergraduate from Macdonald College (now part of McGill University) in Quebec was accused of trying to "poison" his room-mates with a massive dose of *Ascaris* eggs put into food he prepared for them. The grudge was over a not uncommon disagreement over unpaid rent and the accused, who was evicted, had vowed to get even by putting parasites in their food. It was alleged he stole *Ascaris* eggs from a laboratory where he was studying to spike a festive Winter Carnival dinner he prepared for his room-mates. This was in mid-winter and his defense argued that the room-mates could have been infected because of an alleged sewage backup into the kitchen sink or from handling contaminated clothing of the accused. Nevertheless the four room-mates were hospitalized with serious pulmonary symptoms and three others who had shared the food were mildly affected. All of them developed *Ascaris* infestation and passed juvenile worms in stool. One of the room-mates was estimated by doctors to have been infected by as many as 400,000 larvae and had probably suffered permanent lung damage. The accused was charged with intentionally endangering the lives of the victims but eventually the case was dismissed by the presiding judge for lack of sufficient evidence to prove his guilt beyond reasonable doubt.

The adult worms that live in the lumen of the host intestines cause mild symptoms such as abdominal discomfort, and sometimes vomiting and diarrhea, in light infections. In heavy infestations sometimes tangled worms would block the intestines completely and become life threatening. Sometimes adult worms may cause appendicitis when they migrate into the appendix and cause a blockage. They also have been reported to migrate into bile ducts, the liver, occasionally the gall bladder, pancreatic ducts and sometimes into the stomach to be vomited out to the horror of the patient. There have even been reports of *Ascaris* worms crawling out of the nose. However since the heaviest prevalence of ascariasis coincides with low resource settings where under-nutrition is common, it is actually the insidious effects of worm infections aggravating an already precarious public health situation that poses the greatest debilitating effect in poor countries. Treatment is effective with albendazole, mebendazole and ivermectin.

Since broad spectrum antihelminthic medications such as albendazole against human intestinal worms are effective, inexpensive and readily available in developed countries, it is the zoonotic acariids of domestic animals that pose greater health risks in the rich countries. These are *Toxocara canis* in dogs and *T. cati* from cats that cause visceral larva migrans which infect humans when the eggs of these animal parasites are accidentally ingested. The larvae cause much more serious disease than the animal hookworms we discussed before since *Toxocara* larvae could migrate into the eyes, lungs, and other visceral organs. It was reported that as many as 14% of the general US population may have been infected at some point in their lives and in other parts of the world it could be as high as 40% of the general population. A study of 100,000 Irish schoolchildren aged 4–19 showed a prevalence of 9.7% with ocular toxocariasis which is extremely troubling as it is associated with serious vision loss. In the United States, a study in Alabama estimated one case of ocular toxocariasis per 1000 persons in the general population there. Thus toxocariasis remains a fairly serious public health issue in developed countries. Albendazole and mebendazole are effective treatments.

***

Fig. 8.10 Trichuris species
of animals (Source:
Encyclopedia of
Parasitology Ed. Melhorn
2016, Fig. 2)

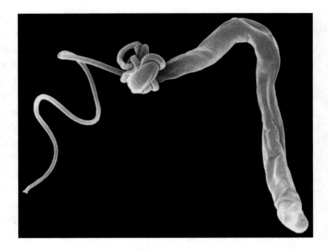

Fig. 8.11 Trichuris species
of animals (Source:
Encyclopedia of
Parasitology Ed. Melhorn
2016, Fig. 3)

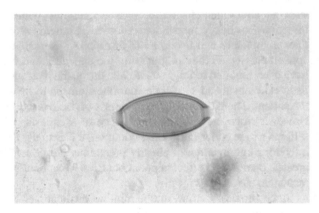

The third member of the intestinal worms that constitutes the triad of soil transmitted nematode parasites in tropical medicine is *Trichuris trichiura*, commonly referred to as the whipworm. The tiny head of the worm joins an elongated slender whip-like body which ends in a thick posterior end that resembles the grip of the whip (Fig. 8.10). The eggs are oval in shape like footballs (that is the American football or the rugby ball) with plugs on both of the narrow ends (Fig. 8.11). The egg is the infectious stage and contamination of food by the eggs from food handlers, direct contamination of hands with infested soil or when human excrement is used as fertilizer and contaminates vegetables are the most likely causes of infection. The egg hatches in the small intestines and develop into adult worms which migrate into the cecum where the head is embedded into the mucosal surface and sometimes the anterior section is threaded into the folds. They are associated with diarrhea and dysentery, and sometimes in heavy infestations in children, they cause prolapsed rectums where part of the rectum would be pushed out of the anus. Currently mebendazole and albendazole are both effective and inexpensive treatment options in mass drug administration (MDA) for trichuriasis as well as the other two species of soil-transmitted intestinal nematode infections, ascariasis and hookworms, and together with triclabendazole, they are vigorously promoted by the WHO.

A different strategy used to combat soil transmitted intestinal parasites, which was not dependent on pharmaceutical treatment, was pursued by my departmental colleague Ricardo Izurieta. The project was to design latrines which incorporated passive solar panels that were integrated into outhouses to raise the temperature of the sewage in the concrete sump below. The "solar latrines" were simple and inexpensive in design, utilized local materials available in poor resource settings, and exploited the tropical sun to raise the temperature high enough to kill even *Ascaris* eggs. It produced safe, odorless, parasite-free organic compost at the end of the composting cycle which could be used as fertilizer. In the latest prototype, Izurieta has improved its capacity to inactivate *Ascaris* by producing a photocatalytic reaction in which solar energy speeds the release of ammonia (when water is added to urea, a widely used fertilizer). This had been successfully field tested in Central America, and is the sort of sustainable and innovative low-tech program, adapted to improve the health of rural communities in tropical countries, that we should hear more about in the media.

∗∗∗

Apart from the triad of intestinal nematodes that infect humans, *Strongyloides stercoralis* remains as another important and widespread parasite in the same category. It is however especially interesting because it exhibits three possible life cycles: a free living cycle, a parasitic cycle with an external infective stage, and finally a completely parasitic auto-infective cycle that never leaves the host body. In the first cycle, *S. stercoralis* rhabditiform larvae passed out from the human host in feces would develop into free living adult male and female worms which live in the soil, copulate and produce eggs. The eggs hatch out in soil and develop into rhabditiform larvae, molt and develop as filariform infective larvae which penetrate the skin the way hookworms do. They would then follow the migratory pathway of migrating into the lungs and then getting swallowed back into the intestines as we have discussed with other nematode life cycles, or they can migrate into the intestines directly via the connective tissues and develop into adults. The second type of cycle is when the rhabditiform larvae excreted in feces will bypass the adult stages and develop directly into the infective filariform larvae and penetrate the skin. From there they will develop as in the first method. In the third auto-infective cycle, the filariform larvae after entering the intestines would develop into adult female worms, mature and the females produce eggs by parthenogenesis (without fertilization) within the intestines (Fig. 8.12). The eggs will hatch out as rhabditiform larvae which will develop into infective filariform larvae in the large intestines. They can migrate into the tissues and eventually re-enter the intestinal lumen and grow into adult female worms and repeat the cycle in the host. In this case an infected individual will have persistent unremitting reinfections even when they are removed from endemic environments. In immunodepressed individuals the filariform larvae

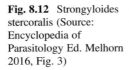

Fig. 8.12 Strongyloides
stercoralis (Source:
Encyclopedia of
Parasitology Ed. Melhorn
2016, Fig. 3)

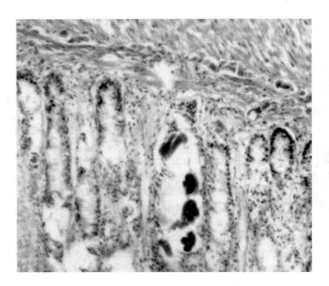

can be disseminated throughout the body and cause hyperinfections. Symptoms include diarrhea, abdominal pain and pulmonary complications. Disseminated hyperinfections in immunocompromized patients include neurological involvement and septicemia and are often fatal. Albendazole and ivermectin are effective to treat strongyloidiasis.

The internal autoinfective cycle within the host body is observed in only one other intestinal nematode parasite, *Capillaria philippinensis*. In that case the natural cycle is found in the intestines of marine fish-eating birds which are the definitive host of the adult worms. Several species of fish are the intermediate hosts. When infected fish is ingested insufficiently cooked by the human host, the worm will burrow into the intestinal mucosa of the person and grow into the adult worm. Eggs released by the adults will usually pass out in feces. Symptoms include nausea, vomiting, diarrhea and weight loss. However their eggs may also hatch into larvae in the intestines and cause autoinfection by developing into adults in the body. Repeated autoinfective cycles would lead to hyperinfection which could be life threatening. Albendazole is an effective pharmaceutical treatment.

Other zoonotic intestinal nematodes include *Anisakis simplex* and related species such as *Phocanema* spp. which have become more prevalent with the increased popularity in cosmopolitan cities worldwide of eating raw fish in sashimi, and also in the Nordic countries and the Netherlands where consuming lightly pickled fish such as herring is common. The third stage infective larvae embedded in the tissue of fish and squid, when consumed uncooked, would penetrate the mucosa of human stomach and intestines and cause abdominal pain, nausea and vomiting. The incidence of infections worldwide is probably underestimated because definitive diagnosis is by endoscopy and physicians would successfully treat with albendazole based only on presumptive diagnosis. Anisakiasis have therefore rarely been

Fig. 8.13 Anisakis
(Source: Encyclopedia of
Parasitology Ed. Melhorn
2016, Fig. 3)

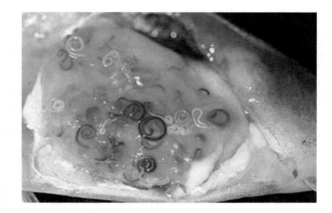

conclusively confirmed by endoscopy and the reported incidence of anisakiasis may only be the tip of the iceberg.

Other factors that contribute to increased incidence of anisakiasis include the "industrialization" of the fishing industry whereby the process of cleaning and degutting of freshly caught fish on the fishing vessels, practiced in the pre-industrialized era, is now entirely bypassed (Fig. 8.13). Modern fishery methods by large modern vessels made it more cost effective to simply store the catch in refrigerated holding tanks until the ships return to shore for the cleaning and evisceration to be done on land. This may take several days, even weeks, during which time the parasite larvae may migrate out of the gut tissues of fish, which is their natural site, into the muscles closer to the surface probably due to the lack oxygen in the deeper tissue of the dead fish. Thus parasite larvae that used to be discarded with the fish viscera would now be found in the flesh. There have also been recent speculations that conservation efforts may have led to increased populations of marine mammals such as seals, porpoises and whales in the oceans, which are the definitive hosts of the parasites. This could contribute to increased numbers of fish becoming intermediate hosts of the parasite larvae which causes anisakiasis. Albendazole is recommended by the CDC for anisakiasis.

Trichostrongyliasis is another intestinal nematode and its occurrence in humans is a result of incidental infection with the infective larvae of *Trichostrongylus colubriformis*, *T. axei* and *T. orientalis*. The most common route of infection is by consuming unwashed vegetables contaminated with parasite larvae from animal manure used as fertilizer. The natural definitive hosts are herbivores such as cattle and the geographic distribution of trichostrongyliasis is worldwide wherever life stock is raised. Treatment with pyrantel pamoate and albendazole are both effective.

∗∗∗

Finally there is the very common, cosmopolitan and ubiquitous intestinal nematode *Enterobius vermicularis* (Fig. 8.14). Known by its common name, the pinworm, it is a frequent infection of children and occasionally in adults taking care of children in institutional settings. It causes itching and irritation in the perianal area

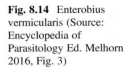
Fig. 8.14 Enterobius vermicularis (Source: Encyclopedia of Parasitology Ed. Melhorn 2016, Fig. 3)

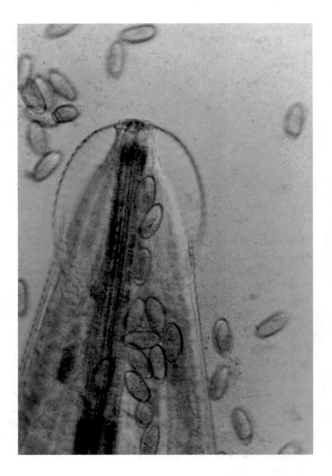

and interrupted sleep in children. In households with children infected with pin-worms, the infectious eggs are often found in bedding, sheets, curtains, and clothing, wherever children scratch and touch. Autoinfection is very common as children would scratch themselves and then put their hands into their mouths. Frequent changing of bed clothes and bedding linen, and washing them in hot water, would be recommended as preventive measures. Albendazole is effective for treatment.

Chapter 9
Tissue Helminths: Snails, Rats, Cobras and Parasites in the Brain

In the 1970s Malaysia's palm oil industry was booming. The country had relied on tin and rubber to fuel its economy when it was still a British colony, but after independence in 1957 it looked around to broaden its economy by adding new products to expand its basket of commodities for export. So the agribusiness widened to include pepper, cocoa, tea, coffee and palm oil. It was the palm oil industry that really took off, as a combination of suitable climate, soil type, cheap labor, good infrastructure and an expanding global market for consumer products all coincided in the 1960s and early 1970s. Palm oil was an inexpensive and versatile ingredient in the manufacture of a constellation of consumer products: cosmetics, soap, shampoo, cooking oil, skin lotion, animal feed, food additive, etc. and it propelled the Malaysian economy into high gear during that period. Therefore new oil palm plantations proliferated rapidly across the coastal flatlands of Peninsular Malaysia.

I had a research project during the early 1970s to study a rat lungworm called *Angiostrongylus malaysiensis*, which can infect the brain of people who consumed salads with vegetables contaminated with their infective larvae. There were outbreaks of human disease in Taiwan, China and several other countries in Southeast Asia in the 1960s by the closely related species *Angiostrongylus cantonensis* (Fig. 9.1). The United States was not spared from this parasite and as recently as April 2017, the Hawaii State Department of Health confirmed an outbreak of six new human cases of *Angiostrongylus cantonensis* infections on the island of Maui and three cases on the Big Island over the previous 3 months. Although no deaths have been reported in this latest outbreak, a handful of cases have been routinely reported annually in Hawaii since the 1960s and two cases of fatalities associated with this infection had occurred in that state since 2007. The parasite is essentially a parasite of rats, which are the natural definitive hosts of the adult worms and humans are incidental hosts. Terrestrial snails belonging to the class Gastropoda and several families of slugs transmit the disease as intermediate hosts in the Asian and South Pacific region. The non-native African giant snail *Achatina fulica* will also serve as

B. H. Kwa, *The Parasite Chronicles*, https://doi.org/10.1007/978-3-319-74923-5_9

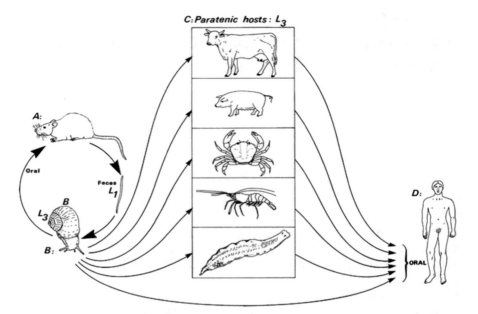

Fig. 9.1 Angiostrongylus cantonensis (Source: Encyclopedia of Parasitology Ed. Melhorn 2016, Fig. 1)

an intermediate host in the region. The parasite continues to spread to other sub-tropical regions and have been reported in Florida since 2013.

During that period when I worked at the University of Malaya, I would regularly drive out with some students from my UM Honors class to one of the many oil palm plantations in the outskirts of Kuala Lumpur to collect land snails and slugs to bring them back to the lab to dissect for the *A. malaysiensis* larvae. The reason why oil palm plantations are such prime spots for *Angiostrongylus* infected snails is that many of the rats in the plantations fed on the snails there (mainly *Macroclamys malayanus* and *M. resplendens*) and became infected with the adult parasites, thus maintaining the parasite life cycle. The habitat in the oil palm plantations was optimal for the propagation of the cycle as there were plenty of discarded palm fruits scattered about which provided food for the rodents, and the large number of palm fronds on the ground from regular trimmings provided ideal nesting places for rats. Similarly a dense population of snails and slugs would graze on the rich organic debris in the shade under the fallen fronds hidden from the tropical sun (Fig. 9.2).

It was an easy matter to overturn the dead palm fronds lying on the ground and to collect the numerous snails and slugs from the ground underneath and sometimes the snails could be found attached to the underside of the fronds as well. The project went on very well as we would always return with a rich harvest of snails and had collected more than sufficient larval parasites to complete the student projects over that year. It was only a couple of years later, when I was advisor to my doctoral student Stephen Ambu, who was then working in the IMR and doing his dissertation

Fig. 9.2 Macrochlamys resplendens, host for Angiostrongylus malaysiensis. Picture by author

research on *A. malaysiensis*, that I happened to talk over coffee with a group of his lab technicians and field collectors. They were very familiar with the oil palm plantations where I had been collecting snail specimens, and they told me how dangerous it could be to collect from under the fallen palm fronds (Fig. 9.3). Apparently several plantation workers harvesting the oil palm fruits had been bitten by cobras which commonly made their nests under the fronds, as the cobras were attracted to those plantations where there were plenty of rats for them to feed on. The cobras were usually very aggressive when they were protecting their eggs or hatchlings, and would fatally attack anyone coming near. Those students and I were very fortunate that we must have collected for snails during the season when the snakes were not protecting their nests, since we were totally unaware of the dangers. After the experience of working with the Australian Tiger Snakes in the past I should have known better and I really must make a mental note to stop doing research on parasites associated in any way with poisonous snakes!

In the rat host, eggs released from the female *Angiostrongylus* worms living in the pulmonary arteries would get trapped in the finer terminal arterioles surrounding the alveoli. The larvae that hatch out would penetrate into the airways and migrate up the trachea and eventually to the pharynx and swallowed into the intestinal tract. There the larvae will pass out in the feces and enter the soil where they will be ingested by the snail or slug intermediate host. They would then develop and molt twice and become infective third stage larvae in the snail. In the natural cycle, when the snail is eaten by the rat, the infective larvae would penetrate the intestines and migrate first to

Fig. 9.3 Palm oil plantation, Sungei Buloh, Malaysia circa 1978. Picture by author

the brain before returning to the venous blood and eventually into the arteries of the lungs where they mature into adult worms.

Human infection occurs in several ways. In instances where consumption of snails is part of the dietary custom, raw or inadequately cooked snails harboring infective larvae are the main source of infection. Occasionally paratenic hosts like freshwater shrimps or crabs may also be infected when they feed on small snails or slugs, and when eaten raw or undercooked are sources of infection as well. Sometimes tiny infected slugs hidden in vegetable leaves could be accidentally ingested in salads. There have been some evidence that infective larvae could be released in the slime trails left on the surface of vegetables by migrating snails, and if not adequately washed would be another source of infection. Once ingested, the infective third stage larvae would penetrate the human intestine and migrate into the brain where they cause meningo-encephalitis. Since humans are not the natural hosts, the worms would not find the necessary physiological markers to find their way back down to the lungs, and would remain in the brain until they die. In studies with *A. cantonensis,* human infections have been reported to last between 2–8 weeks and patients present with symptoms similar to bacterial meningo-encephalitis, such as nausea, vomiting, neck stiffness, and severe headaches. Additionally, the eyes can be affected as a result of inflammation in the region of the brain that affects vision. Human infections by *Angiostrongylus cantonensis* and *A. malaysiensis* are identical although human infections with *A. malaysiensis* is comparatively uncommon and limited to Malaysia. There is no specific treatment for angiostrongyliasis, and treatment is supportive for pain and reduction of inflammation with corticosteroids.

There have been reports of some success with mebendazole and albendazole in limited cases.

There is another species called *Angiostrongylus costaricensis* found in Central America and the Caribbean which infects the intestinal arteries of humans, especially in the ileo-cecal region. There is no brain involvement and it is not life threatening, and the incidence is relatively rare. There have been no proven treatment for *A. costaricensis*.

***

Penang is an island off the northwestern coast of Peninsular Malaysia at the entrance of the Straits of Malacca. It was once a major trading port that is halfway between the Indian Ocean to the west and the South China Sea and the ports of China and Japan to the east. The great navigators in the fourteenth century like Ibn Battuta from Morocco had stopped in Penang on his voyages to Southeast Asia; and Zheng He who had sailed from China contemporaneously in the opposite direction to the coast of East Africa, had also dropped anchor in Penang. It was to be another two centuries later before European navigators like Vasco da Gama reached Asia by sea in the sixteenth century.

I did most of my schooling in my home town Taiping at a school called St. George's, which was run by Catholic missionaries belonging to the La Salle Brothers teaching order. However I had my Sixth Form education (pre-university preparatory classes in the British colonial school system) at St. Xavier's in Penang. This was the first time I lived away from home and I relished the freedom I had, living in rented rooms with two other boys in a city famous for its *Nyonya* cuisine which is a blend of Malay, Chinese, Indian, Thai, Burmese, Javanese, and Sumatran cooking influences. This was the original "fusion" cuisine that the *cognoscenti* kept to themselves and which is only now beginning to become trendy in New York, London, and Sydney. Penang's historical past and its varied immigrant communities who lived there had left their marks as they are reflected in the names of its streets and neighborhoods, such as Arab Street, Armenian Street, Little India, Bangkok Lane, Burmah Road, China Lane, Madras Lane, and Rangoon Road. One of Asia's great leaders of the modern era Sun Yat-sen, the man who established the Republic of China after overthrowing the Manchu Dynasty, had lived at No. 120 Armenian Street.

My student days in Penang had stimulated in me an interest to read old newspaper articles and historical accounts of life in Penang during the pre-war years when it was still a bustling port at the crossroads of the Far East. My father had started out as a shipping clerk in those days at one of those colonial British trading companies headquartered in Penang. I remembered once reading some article in the now defunct broadsheet "The Straits Echo" about the lives of immigrant traders and seamen from Madras (now Chennai in Tamil Nadu, India) who lived in Penang in the 1930s and 40s. The writer reported that he had once seen an Indian man sitting in an alley near the wharf in Georgetown, slowly rolling out with a small stick a long white worm from his ankle, and had wondered what it was. From the description, that could only have been the Guinea Worm, a parasite that belongs to the species

Dracunculus medinensis, which literally means "the little dragon of Medina" (Fig. 9.4). It has been suggested that the symbol of Medicine, the Staff of Asclepius, which shows a "snake" wrapped around a stick, may actually represent the Guinea worm.

This worm infection had been recorded since ancient times, and because of its size and the burning pain involved, is thought to be the "fiery serpent" mentioned in the Old Testament. Several ancient Egyptian texts from the 2nd millennium BCE had mentioned dracunculiasis, and the worm's existence during that period in Egypt was confirmed by finding a calcified male worm in a mummy examined by the Manchester Egyptian Mummy Project. It was also referred to in the Sanskrit book Rig-Veda in India from the fourteenth century BCE. Medical historians found evidence that it was brought to Mesopotamia by prisoners transported from Egypt to Assyria in the seventh century BCE, as it was recorded in a text from the library of King Ashurbanipal at Nineveh. It was later mentioned by authors in the Greco-Roman period such as Galen and Plutarch and by Arab-Persian physicians like Avicenna in the ninth century CE who recorded that he had treated patients with dracunculiasis. The WHO had written an interesting historical account on their website that chronicled this.

Human infections occur with the ingestion of water contaminated by tiny cope-pods (a type of freshwater crustacean) containing infective third stage larvae of *Dracunculus* worms. In the stomach the copepod dies and is digested, releasing the infective larvae which will penetrate the intestines and enter the visceral cavity and retroperitoneal space of the human host. They develop, molt and mature into adult worms and copulate, after which the male worms will die. The females migrate to the

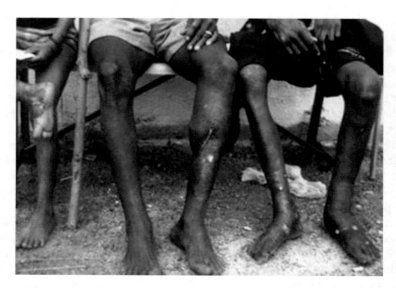

Fig. 9.4 Dracunculus medinensis (Source: Encyclopedia of Parasitology Ed. Melhorn 2016, Fig. 3)

subcutaneous tissues in the lower limbs of the infected person and eventually cause a blister on the skin. When the host comes into contact with water, the female worm ruptures the blister and protrudes out with its distal end and releases the larvae. Copepods in the water will ingest the larvae which will then penetrate into their body cavity, molt twice to become infective third stage larvae, and complete the cycle (Fig. 9.5).

Since no effective pharmaceutical treatment or vaccine is available for Guinea worm, an infected person in impoverished communities would hold the emerging end of the worm and slowly pull it out. To do this they would patiently and gently roll it around a twig or stick, constantly sprinkling water while doing so over a period of several hours or even days. As the worm is sometimes a meter long and embedded in the tissues of the leg, any abrupt attempt to pull it out quickly will rupture the worm while a large part of it would still be left within the leg. The dead or dying worm within the tissues would likely cause an anaphylactic reaction as well as result in secondary bacterial infection which would only aggravate the infection.

The Carter Foundation has made the eradication of Guinea Worm a priority and they have focused on prevention by promoting the use of nylon netting to filter drinking water and to encourage people to only drink filtered water. Educational

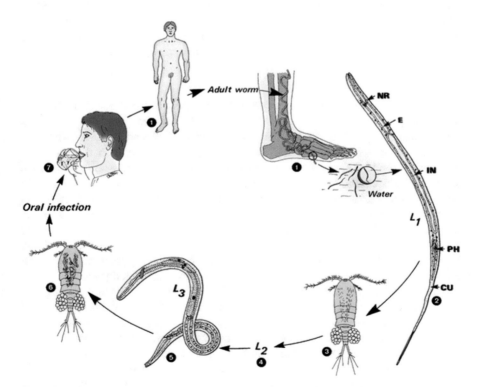

Fig. 9.5 *Dracunculus medinensis* (Source: Encyclopedia of Parasitology Ed. Melhorn 2016, Fig. 1)

programs also emphasized the prevention of infected persons from physically entering water sources and by designing and building wells that prevent entry into the water source. The application of pesticides such as Abate to kill the copepods is used in open waters such as ponds and streams. Guinea worm infections used to be prevalent in Africa, the Middle East and in the Indian Subcontinent. The Carter Foundation has reported remarkable success in eradication and has reduced the number of cases from 3.5 million when they started, to just 22 cases worldwide by 2015.

Recently in 2016 however there had been a mysterious outbreak of the disease in dogs in Chad, and this was a serious concern because it might reintroduce the disease to humans in regions where it had previously been eradicated. Dogs are unlikely to be infected from ingesting copepods in contaminated waters because the lapping motion they use in drinking scares away copepods. Initial research indicated that they were instead probably infected from eating discarded fish entrails of gutted fish which had fed on copepods and which contain the parasite larvae. The dogs would then contaminate the water by reintroducing infective larvae into water which had previously been cleared of infected copepods, and restart the human Guinea worm cycle. An emergency response included paying the local population to identify infected dogs and tie them to prevent them from contaminating water sources. Efforts are also being made to encourage villagers to bury the discarded fish entrails to prevent dogs from eating them. There are also trials to test if treatment with dog heartworm medication such as ivermectin would have an effect on the Guinea worm. Ongoing results have so far been encouraging.

***

Contaminated meat products, especially uninspected game meat, harbor the nematode parasite *Trichinella spiralis* and the disease they cause is known as trichinellosis. Up to the 1940s pork products in the US had been known to be important sources of infection, and the US Public Health Service had reported an average of around 400 cases annually then. In North America it is no longer a food safety issue in commercial meat products with the advent of modern pig raising methods, freezing of meat in modern slaughter houses, strict meat inspection and efficient transportation in refrigerated trucks by the commercial meat industry. Home freezing of pork is also safer nowadays since *Trichinella spiralis* is killed by temperatures below -15 °C for 20 days. However other species such as *T. pseudospiralis* are not killed by freezing, and occasionally there are reports of hunters becoming infected after consuming wild game they killed (Fig. 9.6).

Human infections with *Trichinella spiralis* begin by the ingestion of encysted parasite larvae in infected meat. The larval cysts will be digested by host gastric acid and pepsin in the stomach and the liberated larvae would develop and mature into adult worms in the small intestines. After copulation, female worms would release larvae that penetrate the intestinal mucosa, migrate and invade striated muscles where they would encyst. Humans are "dead end hosts" since they do not pass on the infection. In sylvatic cycles in nature, wild animals such as bears, wild boars and rodents are the natural hosts and rodents are the main vector for transmission as they

Fig. 9.6 Trichinella spiralis and related species (Source: Encyclopedia of Parasitology Ed. Melhorn 2016, Fig. 4)

are sufficiently low in the food chain to be regularly preyed upon by the larger omnivores and the larval cysts in their muscles will propagate the disease. In the domestic cycle, in small farms outside the commercial meat industry, a similar cycle among pigs and rodents maintain the cycle, when pigs have access to carcasses of infected mice (Fig. 9.7).

Human disease involves an initial phase associated with gastrointestinal symptoms caused by the infective larvae being released from the cysts in the stomach and migration to the intestinal tract and their development into adult worms. These symptoms last 1–2 days and include abdominal pain, nausea, vomiting, and diarrhea. Subsequent symptoms are caused by the migrating larvae produced by the mature female worms and these usually begin within 2 weeks and can continue for several weeks depending on the worm load and numbers of larvae produced. These tend to be severe immuno-pathological reactions to migrating parasites in the muscles and may include myalgia, fever, skin rash, facial edema particularly swelling around the eyes, headache, chills, cough and general fatigue and elevated eosinophilia. In heavy infestations, breathing and heart problems such as myocarditis, encephalitis, and thromboembolic disease may occur and they can be life threatening. Albendazole and mebendazole are both effective in clinical treatment of trichinellosis.

∗∗∗

Finally there is one other nematode infection in humans, gnathostomiasis, that is mostly associated with consuming raw or undercooked fish and other aquatic creatures like frogs and eels. The parasite is *Gnathostoma spinigerum* and originally it was found in tropical areas of Southeast Asia, but have now been reported in many

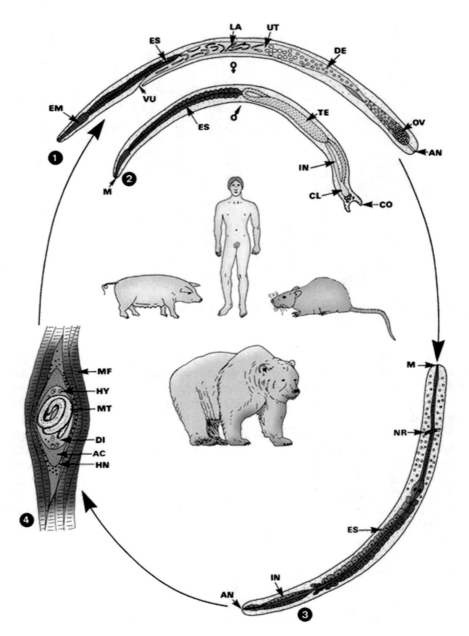

Fig. 9.7 Trichinella spiralis and related species (Source: Encyclopedia of Parasitology Ed. Melhorn 2016, Fig. 1)

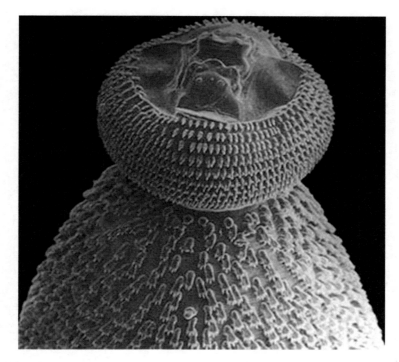

Fig. 9.8 Gnathostoma spinigerum (Source: Encyclopedia of Parasitology Ed. Melhorn 2016, Fig. 2)

parts of Central and South America as well (Fig. 9.8). The adult worms live encysted in the stomach wall of dogs and cats, both domestic and feral. Eggs are passed via a canal from the cyst into the lumen of the stomach and are eventually passed out in feces. In water, the first stage larvae hatch out and are eaten by copepod intermediate hosts within which the second stage larvae will develop. These in turn will be eaten by fish and amphibians such as frogs or toads where they develop into third stage larvae. They would remain as third stage larvae when ingested by paratenic hosts such as snakes and birds, and will not develop any further. Only when ingested by the definitive cat or dog hosts will they develop and mature into the adult worms.

Humans are incidental hosts and usually acquire the parasite by eating raw or undercooked fish. The infective third stage larvae penetrate from the gastrointestinal tract into the tissues and will migrate into tissues such as under the skin and occasionally into the visceral organs such as the liver. They sometimes migrate into the eye, resulting in vision loss or blindness. More rarely their migrations involve the nerves, spinal cord, or brain, resulting in severe complications that include nerve pain, paralysis, coma and even death. There have been reports of successful clinical treatment with albendazole and ivermectin. Avoidance of consuming raw fish in general is advisable.

Chapter 10
Intestinal Protozoa: Please Pass the Stool

Among developed countries of the West, towards the end of the twentieth century, there was a widespread belief that unlike countries in the "Third World" where the drinking water was generally contaminated and the public health infrastructure was suspect, you could always rely on the fact that it was safe to drink water from the tap here. I remember that as one who grew up in the Third World (a common designation for Malaysia before the era of politically correct euphemisms) I was shocked to see my Australian college room-mate drink straight from the tap during my freshman undergraduate year in 1965 at the Australian National University (ANU) in Canberra. In my childhood, Malaysians were taught to only drink boiled water as a general rule and the idea of drinking "raw" water straight from the tap was to invite the risk of tummy upsets and worse. When I had mentioned this to my fellow students at ANU, some of the less sensitive Aussies would say that "only the wogs have to do that"—that is, only the great unwashed in backward countries have to boil their water. This attitude was still prevalent in the 1960s when the last vestiges of the White Australia policy had yet to be rescinded and Arthur Calwell, who had previously declared in parliament when he was Immigration Minister, that "two Wongs do not make a White" was then leader of the Australian Labor Party.

The outbreak of cryptosporidiosis in the city of Milwaukee in 1993 was to forever change attitudes about the safety of our municipal water supply much more deeply than most people realize. Historians and anthropologists will probably trace it back to that year to record the start of the trend of seeing the ubiquitous plastic bottle of potable water in the hands of practically everyone in the developed world nowadays. Milwaukee was such a seminal event for a number of reasons. First of all, it was a massive outbreak in which more than 400,000 people were estimated to have become ill with diarrhea in a single event. Second, it was a result of people drinking piped water supplied from a municipal source which had followed modern standards of safety associated with treatment procedures of flocculation, sedimentation, filtration and disinfection with chlorination and properly tested for turbidity. Third, it had been a "silent" epidemic which involved a quarter of the inhabitants of a major city but could have been largely unreported until the media played up the story and the

B. H. Kwa, *The Parasite Chronicles*, https://doi.org/10.1007/978-3-319-74923-5_10

outbreak became known worldwide. An earlier smaller outbreak had occurred in 1987 in Carrollton, Georgia in the US but because it had not received the same publicity, it did not have a similar impact. The publicity generated by the Milwaukee epidemic had however triggered similar investigations in other developed countries where similar outbreaks were soon reported in Britain and Canada in communities where strict municipal water standards had also been met.

Interestingly, on the morning of April 5, 1993 the first indication of a major outbreak of diarrheal illness in Milwaukee was discovered by anecdotal information that many of the local pharmacies were selling out of over-the-counter anti-diarrheal medications. Concurrently the Milwaukee Water Works had received numerous complaints from residents about the aesthetic quality of the water coming out of their tap. The water was cloudy due to high turbidity and had a bad taste and odd smell. This led to the start of investigations by the City of Milwaukee Health Department which ultimately reported the massive outbreak.

The culprit was a protozoan parasite called *Cryptosporidium* of which there are two species which infect humans. One species is *Cryptosporidium hominis* which is an exclusively human to human parasite and the other is *C. parvum* which is transmitted among ruminants, mainly cattle, but is also infective to humans. Many of the large municipal outbreaks were caused by *C. parvum* and had been associated with contamination of municipal water sources by runoffs from dairy farms during periods of heavy rainstorms when the treatment facilities were overwhelmed by excessive contamination. The parasite is a single celled protozoan 4–6 μm in size that belongs to the phylum Apicomplexa and had been first described in mice in 1907 (Fig. 10.1). The parasite was only first known to be a human pathogen in 1976 when it was described to infect the epithelial lining of the human small intestine but remained an obscure human parasite which was seldom reported. In healthy individuals with intact immune systems, cryptosporidiosis causes a self-limiting intestinal illness causing diarrhea which is resolved within 9–15 days. The clinical presentation includes watery diarrhea containing mucus but rarely blood or leucocytes. It may also include cramps, nausea and abdominal pain and sometimes mild fever but does not cause death. However with the emergence of the AIDS epidemic it soon gained prominence as an opportunistic infection associated with immunocompromized individuals, where it was recognized to be a life threatening causative agent of unremitting diarrhea. This established the importance of *Cryptosporidium* as a significant human pathogen and the large municipal outbreaks from chlorinated water sources merely enhanced its prominence in the clinical literature.

The infective stage is the oocyst which leaves the host upon defecation and which is protected by an oocyst wall with inner and outer shells composed of a protein-lipid-carbohydrate matrix which is resistant to chlorination (Fig. 10.2). Within each oocyst are four motile infective sporozoites which are released when the oocyst is ingested by an appropriate host. The excystation process occurs in the gastrointestinal tract and is mediated by parasite-derived enzymes and proteins which are stimulated to be secreted when host factors including temperature, pH, carbon dioxide, pancreatic enzymes and bile salts are present. Each of the sporozoite is capable of attachment to epithelial cells of the host intestinal endothelium and upon

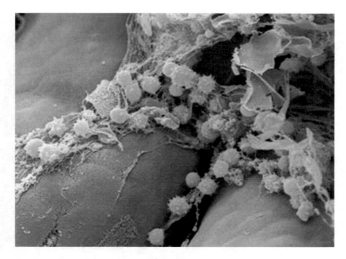

Fig. 10.1 *Cryptosporidium* species (Source: Encyclopedia of Parasitology Ed. Melhorn 2016, Fig. 5)

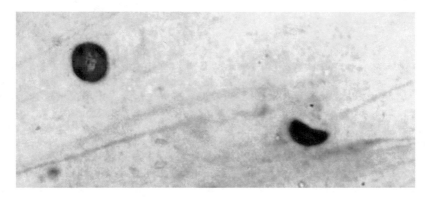

Fig. 10.2 *Cryptosporidium* species (Source: Encyclopedia of Parasitology Ed. Melhorn 2016, Fig. 3)

entrance into the cell becomes encapsulated by a parasitophorous vacuole common to apicomplexans (Fig. 10.3). What make *Cryptosporidium* unique among apicomplexans is that the parasitophorous vacuole of *Cryptosporidium* is intracellular *but extracytoplasmic*, and a special feeder organelle is formed to absorb nutrients from the host cell (Fig. 10.4). This feeding stage is now called a trophozoite and it will proceed to reproduce asexually to form meronts. Two types of meronts are formed: Type I meronts will continue to reproduce asexually, producing up to eight infective merozoites from each meront which will invade surrounding host intestinal cells. This stage causes the diarrhea associated with cryptosporidiosis. A second Type II meront proceeds to produce sexual stages called microgametocytes (male)

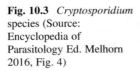

Fig. 10.3 *Cryptosporidium*
species (Source:
Encyclopedia of
Parasitology Ed. Melhorn
2016, Fig. 4)

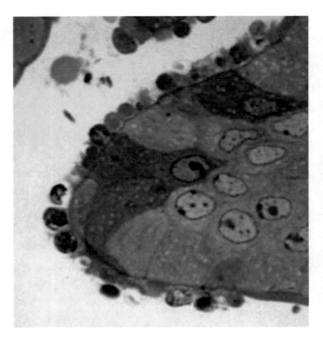

and macrogametocytes (female). These will initiate the sexual reproductive cycle, and fertilization between the microgamete and macrogamete results in the production of thin walled and thick walled oocysts which are released into the intestinal lumen. Thin walled oocysts travel down the gastrointestinal tract and will infect new regions of the intestinal lining further down. However, thick walled oocysts are shed with the feces into the outside environment where it is immediately infectious but will remain resistant to harsh external conditions to await infection of a new host.

For immunologically healthy individuals infected with cryptosporidiosis, the diarrhea mostly is allowed to resolve by itself and patients are usually only given plenty of fluids to prevent dehydration. Currently the only FDA approved medication to treat diarrhea caused by cryptosporidiosis is nitazoxanide, but it is not effective in immunologically compromised individuals. Those with HIV infection have to be treated with antiretroviral medications to resolve the underlying immunosuppression. In cases of known municipal outbreaks, boiling of water is encouraged and otherwise drinking bottled water is indicated. Although newer more effective water treatment methods such as ultraviolet radiation have been introduced in some water treatment plants, they are still relatively expensive to retrofit into existing facilities and are therefore not widespread in implementation. During outdoor activities such as camping or hiking, water from lakes and streams have to be boiled or filtered with portable filters with pore size of 1 μm or smaller, or labeled NSF 53 or NSF 58. Thus cryptosporidiosis is still a major problem of developed countries since current municipal water treatment methods to completely eliminate

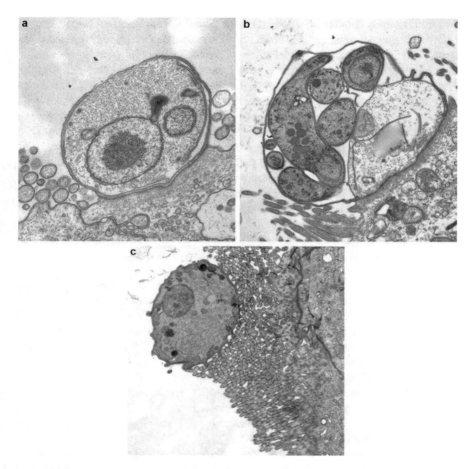

Fig. 10.4 *Cryptosporidium* species (Source: Encyclopedia of Parasitology Ed. Melhorn 2016, Fig. 6)

the parasite are not totally reliable, and an effective pharmaceutical drug for its clinical treatment is still unsatisfactory.

***

Most discussions of environmental outbreaks of waterborne parasitic infections, such as those caused by *Cryptosporidium*, include another parasite called *Giardia lamblia*. Because *Giardia* cysts at low temperatures are fairly resistant to chlorination, it poses a similar challenge to public health measures in North America since the entire infrastructure in municipal water treatment developed over the last century had been predominantly dependent on chlorination. In the U.S. sporadic local outbreaks had been reported even from upscale ski resorts. For example in November 1981 an outbreak of giardiasis occurred in Aspen Highlands, a popular ski resort in Colorado where a group of visitors were infected from drinking municipal tap water.

In some European countries, water treatment plants have introduced other methods such as ultra violet radiation, ozone and reverse osmosis in their treatment and those were more effective in the disinfection of protozoa such as *Giardia* as well as *Cryptosporidium*. Unfortunately retrofitting these treatment equipment and machinery into current treatment facilities at the scale required proved expensive and relatively uncommon in the US.

Giardiasis is sometimes sensationally called "beaver fever" in the American media but the catchy name is not altogether very accurate. Although many wild mammals are carriers and may contaminate natural water sources such as rivers, lakes and mountain streams and thus pose a risk to campers and hikers who drink from them, beavers are not the major carriers. Furthermore fever is not associated with giardiasis. Fortunately boiling or filtering the water is effective in removing the *Giardia* cysts and many effective portable water filters are on the market for outdoor activities. There is also much evidence that humans and their pets contribute as much in contaminating pristine wilderness environments as the wildlife, probably more so as increased tourism in wilderness areas is putting increased stress on the natural environment.

Giardia lamblia is a cause of diarrhea with greasy stools that tend to float, accompanied by symptoms that can include severe abdominal cramps, nausea and vomiting. Dehydration can be a serious complication with extended intermittent episodes of diarrhea, and complications include weight loss and failure to absorb fat, lactose, vitamin A and vitamin B12. It is a very common waterborne infection throughout the world especially in low resource settings. Reports of tourists visiting places as different as St. Petersburg Russia, Cuba and Morocco becoming infected with giardiasis from drinking water from the tap, are becoming commonplace as more adventurous travel to newer destinations increases worldwide. Fortunately current effective pharmaceutical treatment options include metronidazole, tinidazole, and nitazoxanide.

The infective stage is a cyst, around 8–14 μm in size, which is fairly resistant to external environmental conditions. When ingested, the *Giardia* cyst will be exposed to the acidic environment of the stomach and be stimulated to excyst in the duodenum and release two flagellated trophozoites. Under the microscope, each trophozoite is oval in appearance with two prominent nuclei, and they always look comical and remind me of those coin operated oval binocular viewers you see at tourist sites. The ventral side of the trophozoite is modified into a shallow disc which acts as a "sucker" and is used to attach the trophozoite to the mucosal surface of the intestine (Fig. 10.5). Each trophozoite has four pairs of flagella which allow the trophozoites to be motile and able to migrate to new attachment sites on the mucosal surface of the intestine. The trophozoites reproduce by binary division but upon exposure to bile salts some of them will form cysts in the jejunum and be passed out of the gut with the feces to infect new hosts and complete the life cycle (Fig. 10.6).

Among old school parasitologists who examine parasite cysts, oocysts, and ova under the microscope to make diagnostic identifications, the proper preservation of specimens are of the utmost importance. This cannot be overemphasized because very often the microscopist depends on accurate observation of fine details such as

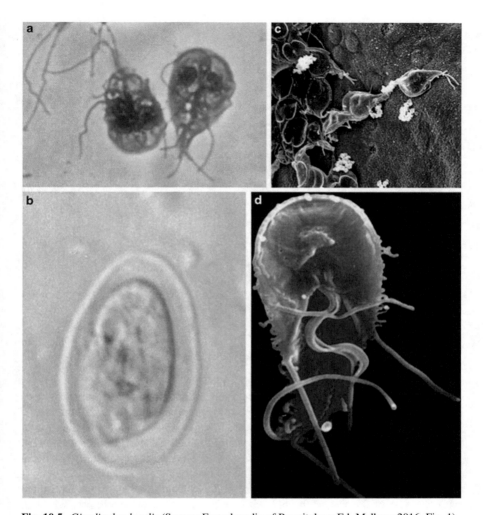

Fig. 10.5 *Giardia duodenalis* (Source: Encyclopedia of Parasitology Ed. Melhorn 2016, Fig. 1)

the size and shape of the cyst, number of nuclei within the cyst, shape and position of the nuclei, etc. for their identification. This is an arcane art as much as a science since it depends so much on the experience and skill of the microscopist as they are working at the very limits of the traditional light microscope, using esoteric methods such as oil immersion lenses to push the limits of magnification.

Furthermore they work with feces, into which all the intestinal parasites release their cysts and ova into the outside environment. Thus the very medium they need to find the parasite specimens and identify them with precision is mixed up with all kinds of debris that flows from the gastrointestinal tract. So the first essential step is to chemically "fix" the parasite cysts and eggs as quickly as possible in a state that preserves their internal structural integrity, since those structures within have to be

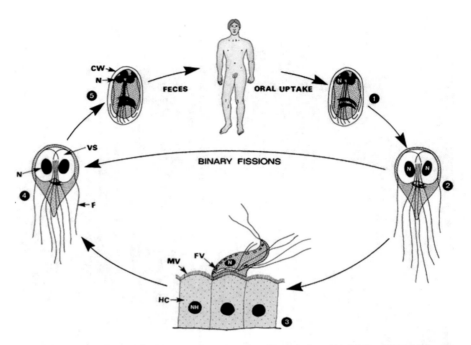

Fig. 10.6 *Giardia duodenalis* (Source: Encyclopedia of Parasitology Ed. Melhorn 2016, Fig. 2)

stained with dyes to reveal their micro-structures and used later for microscopic identification. Thus the microscopist-parasitologist not only needs to find the needle in an extremely dirty (and smelly) haystack, they also need to identify the type of needle they find by examining the fine inscriptions on the needle, as it were. That perfect fixative is therefore the *sine qua non* in the microscopic alchemy of that generation of diagnostic parasitologists.

Soon after I started working at the University of South Florida, I met Don Price who was one of the last of the old school diagnostic lab parasitologists who had dedicated their careers to this arcane science. I had mentioned him in an earlier chapter where I described how he discovered that a group of bridge-playing ladies in Sarasota, Florida were being infected by the Rat Tapeworm from eating snacks contaminated with rat droppings. Don Price was a perfectionist and great stickler to detail. His exacting laboratory discipline required that, and he had a personality to match the demands of his craft. While Don was a very strict and demanding martinet in the lab, and could be downright ornery when crossed, he also had his idiosyncra-sies and a wicked sense of humor. He delighted in wearing sarongs for example, having acquired the habit after living for a few years doing research in Malaysia early in his career in parasitology. He had lamented to me that he could never find sarongs for men (usually printed in broad checked patterns) in the United States but could only find sarongs with flowery batik patterns for women in specialist bou-tiques. So the next time we had visited Malaysia, my wife had bought him a couple of "manly" sarongs as a gift and he was absolutely delighted and wore them proudly on the beach on the way to his morning swim during our next expedition to St. Croix, to the curious stares of the local inhabitants. During that period in the 1990s we were testing parasite collection kits developed by Don for rapid fixation and preservation of protozoan cysts and helminth ova. We had a small grant that allowed us to travel to several locations in the Caribbean and Central America for field testing and that was when I discovered his eccentric and mischievous side.

Before we set out on the first of our many trips we had experimented with various fixatives and worked out the sequence and times of the various procedures in the laboratory at USF under controlled conditions, before we were satisfied with how robust the prototype kit would work under field conditions during the upcoming expeditions. During this period a couple of my graduate students would work with us on those experiments and because the kits were intended to collect fresh fecal samples, we needed fresh human stool samples for the experiments. The first morning Don walked in and demanded to know which of us had brought the required specimen. None of us had been told and so no one had brought in a fecal sample. One of my male students "JS" had been courting a pretty female grad student "D" at the time. As she had not arrived yet that morning, we all conspired to tell her when she arrived that Dr. Price had expected each member of the research team be able to produce a fecal sample on demand and that we had drawn lots and she had been designated to be that person that morning. We all tried our best to keep a straight face when Don announced the demand and sternly passed the collection vial to her and poor "D" took the vial in bewilderment before we all collapsed in laughter. I think I agreed to be the first guinea pig that day to spare her the embarrassment. However despite the smelly and decidedly unromantic circumstances of the parasitology lab, I am happy to report that "JS" and "D" are happily married with a growing family and both are very successful MDs brilliantly pursuing their respective medical specialties in Florida.

More modern molecular technologies that depend on identifying parasite DNA are nowadays more accurate, quicker, and when made into a handy portable kit, a robust and dependable method. They have the added advantage of not having to be dependent on the experience and skill of a microscopist, but are currently still too expensive to be widely used in low resource settings. Furthermore for research purposes, only the old school methods of specimen preservation would allow you to archive a physical collection of parasite specimens for future comparisons to other species. I must confess that I personally find working with fecal specimens absolutely the least attractive aspect of parasitology and am forever grateful when molecular methods supplanted the need to be mucking about with that stuff, no pun intended. However, sometimes those "shitty expeditions" as my students called them, had their lighter moments.

While we were in Belize on one of these expeditions, we had set up a collection kiosk in one of the local public health clinics and took turns manning the different stations. We would explain our project to the patients in the waiting room and would persuade them to participate in our parasite survey. One of us would pass out the informed consent forms, collection vials and would instruct the volunteers the proper procedure to collect the fecal sample in the vials. One of my graduate students in the team was a young British medical graduate doing his MPH in our College of Public Health who was well spoken with a nice posh English accent. I will call him Dr. T. On that occasion the volunteer was a grizzled and rather rough looking character who was either a fisherman or dock worker and was having trouble understanding what he was supposed to do. Dr. T would delicately explain in his prim and proper manner that he should "place his stool" into the vial with the "wooden spatula" provided. The Belizean man was getting frustrated and irritated with the instructions delivered in the clipped Oxbridge accent which he could not understand, and one of my American students was about to walk over to explain when it suddenly dawned on the man what he was supposed to do. "Oh. . . you want me to put my **shit** into the bottle with this wooden stick! Why don't you just say so" he exclaimed in a very loud voice. Our entire team and all the Belizean nurses and med techs in the entire clinic could not help hearing it and we all burst out and rolled over in laughter!

***

On June 10, 2000 more than 50 attendees at a catered wedding reception in Philadelphia became sick with diarrhea and gastrointestinal illness, including the bride and groom and a majority of the guests. An investigation conducted by the Philadelphia Department of Public Health, with collaborators from the CDC and FDA, later established that they had been infected with *Cyclospora cayetanensis* from raspberries in the wedding cake. Furthermore the raspberries were linked to suppliers from Guatemala and this particular outbreak was part of a series of *Cyclospora* outbreaks in the U.S. which had occurred since 1995 which had implicated strawberries from Guatemala, either imported fresh or frozen. As the supply of food products have become globalized, parasitic diseases once thought to be rare in some parts of the rich post-industrial world, have started to occur with increasing frequency. However this should not be an indictment of globalization which is generally beneficial for both the developed economies which benefit from the availability of global products at reasonable prices, and emerging economies which are able to find new markets for their products. However, an enlightened global approach to public health measures is needed to respond to a globalized chain of imported goods, including food products. Indeed the response of the U.S. to the *Cyclospora* outbreaks could be cited as an exemplary global collaboration that effectively dealt with the problem. Once the source of the contamination was identified, and it was only limited to a few specific individual farms in Guatemala, the Guatemalan government cooperated with the FDA and the berry industry to improve farming and exporting practices for raspberries. Only farms that met certain standards, which were reviewed and updated annually, have been allowed to export fresh raspberries to the United States after that. These strict regulations have since been implemented successfully and had resulted in no new outbreaks.

Infections of *Cyclospora cayetanensis* occur through ingestion of sporulated oocysts from feces contaminated soil and vegetables and fruits grown close to the soil such as lettuces and raspberries (Fig. 10.7). When the oocysts are passed out

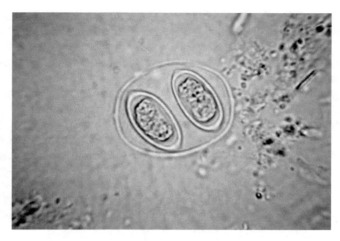

Fig. 10.7 *Cyclospora cayetanensis* (Source: Encyclopedia of Parasitology Ed. Melhorn 2016, Fig. 2)

with feces they are not immediately infectious and require several days to several weeks of "sporulation" to develop the four sporozoites within each oocyst. This variation in the time required for sporulation is thought to depend on environmental conditions such as temperature, humidity and rainfall. Since freshly passed oocysts are not infectious, person-to-person infections do not occur and human infections are mainly associated with ingestion of contaminated uncooked food such as fresh raspberries and salads. This differentiates it from cryptosporidiosis we had seen earlier.

Ingested oocysts will excyst in the small intestines and the infectious sporozoites enter the epithelial lining and invade the cells and undergo the asexual reproductive phase. The infected intestinal cells will release Type I meronts that reinfect adjacent cells, and this phase results in the gastrointestinal illness and diarrhea associated with cyclosporiasis. Some of the meronts develop as Type II meronts and these initiate the sexual reproductive phase, with development of microgametocytes and macrogametocytes which result in fertilization between microgametes and macrogametes, and the release of oocysts via the host feces. The oocysts then undergo the sporulation process in the external environment until they are fully infectious and remain viable they are ingested by the next person. Treatment of cyclosporiasis is successful with Trimethoprim/sulfamethoxazole (TMP/SMX), which is also known by the trade names Bactrim, Septra or Cotrim.

***

For travelers, there is a much more serious waterborne protozoal parasite, *Entamoeba histolytica*, which causes amebic dysentery in nearly all parts of the tropics worldwide. Unlike the other parasites mentioned earlier in this chapter which are not tissue invasive, *E. histolytica* would invade the intestinal lining and cause ulceration in the sub-mucosal layers which will result in bloody diarrhea including inflammatory colitis (Fig. 10.8). They may even cause perforations that penetrate through the intestinal wall which will result in peritonitis. In some cases the amebae could enter the bloodstream and be carried to secondary sites in the liver where it will cause liver abscess (Fig. 10.9). Even more rarely, secondary sites of infections such as the lungs and brain have been reported.

On August 16, 1933 and for several weeks after that, there was a large outbreak of amebic dysentery during the Chicago World's Fair, which was up to that point the biggest reported epidemic in the United States caused by *E. histolytica*. As amebic dysentery is usually reported in the tropics, initially the doctors in charge of the investigation had not been looking for this disease and it was misdiagnosed as

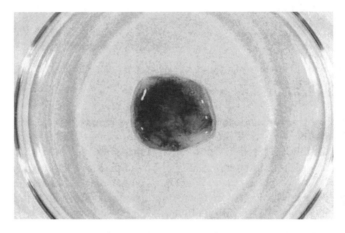

Fig. 10.8 *Entamoeba histolytica* (Source: Encyclopedia of Parasitology Ed. Melhorn 2016, Fig. 6)

Fig. 10.9 *Entamoeba histolytica* (Source: Encyclopedia of Parasitology Ed. Melhorn 2016, Fig. 7)

appendicitis, bacterial colitis and so on. All told, 721 clinical cases from 206 cities were eventually traced back to a visit to the World's Fair in Chicago that summer. As reported in an editorial of the American Journal of Public Health in 1934, a total of 1049 carriers of *E. histolytica* were eventually identified in the city of Chicago. According to the same editorial, it was later discovered that the source of the contamination and 94% of the cases detected came from two hotels which had shared a water tank on the roof of one of the hotels. Unfortunately the water was also used for cooling the hotels' air-conditioning system before being pumped to the tank on the roof meant to store drinking water. The water pipes were mainly made of steel as were all the sewage pipes, which had been the standard practice then, and some of the sewer lines were so badly corroded that it was reported that it was possible to push "a five cent fork through the main pipe". In some places the holes were simply blocked with makeshift wooden plugs which had rotted through and were leaking badly. Unfortunately some of the sewage pipes, which had carried some 62% of the hotels' load, had "passed directly above the tank in which water was refrigerated for the dining rooms and the floors." The account gets even more amazing! "Two four inch sewer pipes intended for overflow emptied into a cement trough under the laundry tubs. All leakage was carried to sumps and forced to the level of the street sewer. . . (and) following two rains on June 29 and July 2, the street sewer broke, flooding the basement of Hotel C and running directly into the ice storage house."

It is important to realize the historical context in which this outbreak had occurred. It was during the Great Depression and Chicago city government had been bankrupted by the previous administration. There were overlapping responsibilities of numerous city agencies such as the "Building Department, Fire Department, License Department, Department of Gas and Electricity, Smoke Inspection Department, Department for the Inspection of Steam Boilers, Department of Public Works and Board of Health" which all had a shortage of inspectors, and routine inspections had been neglected. In-house hotel engineers were "in the habit of making repairs and additions without notifying the city" and all these errors taken together formed the perfect storm when the sudden influx of visitors to the World's Fair overwhelmed the inadequate infrastructure of 1933 Chicago. There were also hints of a cover-up as the 1934 editorial stated that "The health authorities of Chicago have been blamed severely for suppression of the news and it has been charged that it was done in order not to scare visitors away from the Exposition."

The parasite responsible for amebic dysentery, *Entamoeba histolytica,* is a protozoa and human infections occur when its cyst is ingested. The cysts of *E. histolytica* are passed in the feces and they can survive up to several weeks in the external environment under moist conditions such as in soil in the humid tropics. Cysts of *E. histolytica* have four distinctive nuclei and are protected by a thin cyst wall (Fig. 10.10). During severe diarrheal episodes trophozoites could sometimes be found extruded in fecal samples, but these do not survive outside the body and do not play a role in the transmission of the disease. Excystation occurs in the small intestines where the trophozoites are released from the cyst and then would migrate to the large intestines. Trophozoites of *E. histolytica,* in the majority of cases, will

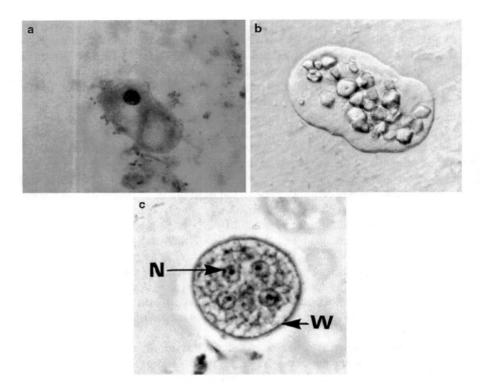

Fig. 10.10 *Entamoeba histolytica* (Source: Encyclopedia of Parasitology Ed. Melhorn 2016, Fig. 4)

live in the lumen of the large intestines feeding on bacteria and dividing by binary fission (Fig. 10.11). Under certain circumstances, not yet fully understood, they will become tissue invasive and enter the intestinal mucosa. The current consensus suggests that human genetic variability in susceptibility, as well as genetic variability of different parasite strains, all play a role in determining whether tissue invasion and pathology occur. Metronidazole (Flagyl) and tinidazole (Tindamax, Fasigyn) are effective for treatment of amebiasis.

There are species such as *Entamoeba dispar* and *E. moshkovskii* which are morphologically indistinguishable from *E. histolytica* but live as benign commensals in the lumen of the large intestine without causing any illness or pathology. They are thought to be responsible for some of the misidentification of *E. histolytica* since they are usually distinguished from *E. histolytica* only by the absence of human erythrocytes in their cytoplasm. Thus you can see how ambiguous current methods of microscopic identification are since often any *E. histolytica*—like parasites would usually be categorized as *E. dispar* if no red blood cells are seen. Molecular methods such as polymerase chain reaction (PCR) are the only reliable techniques for differentiating among these species, but while useful for epidemiological studies they are currently not routinely used for diagnostic purposes due to the cost.

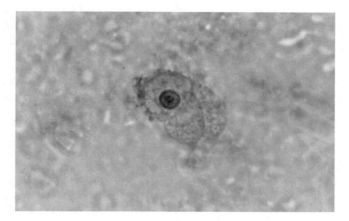

Fig. 10.11 *Entamoeba histolytica* (Source: Encyclopedia of Parasitology Ed. Melhorn 2016, Fig. 3)

Several other species of amebae living essentially in the same environment in the human gut include *Entamoeba coli*, *E. hartmani* and *E. polecki*. These do not cause any illness and appear to live as commensals in the gut without affecting the health of the human host. They are also more easily differentiated from *E. histolytica* under routine laboratory examination.

Chapter 11
Tissue Protozoa: That Cat Will Really Drive Me Crazy One of These Days

In earlier chapters we had shown how parasites have been able to manipulate their hosts to the advantage of the parasite. For example, we have seen how the tapeworm *Schistocephalus solidus* affects their fish host in such a way that the fish becomes more likely to be caught and eaten by birds, thus favoring the tapeworm to complete its life cycle in the bird. We have also discussed how the liver fluke *Dicrocoelium dendriticum* manipulated the behavior of the ant in such a way that enables both to be more likely ingested by sheep which happen to be the final host of that parasite. Those examples however only involved animal hosts, and the idea that animal hosts may have been unwittingly manipulated by parasites is relatively easily accepted without creating much of a fuss.

However a recent theory that human hosts may similarly be manipulated by the tissue protozoan parasite *Toxoplasma gondii* is such a disturbing idea that it became somewhat of an instant *cause célèbre* for the public when it was first reported. What makes it especially controversial is the proposition that it is human behavior that is being manipulated by *T. gondii,* and that evoked all kinds of very deep seated fears of an invisible creature being able to cause us to lose our "Free Choice"—in other words it has the power to turn us into zombies controlled by the parasite! Furthermore the idea that this insidious parasite has even recruited our much loved household pet—the cat—in its devilish quest to subvert our mind is just too scary to contemplate.

T. gondii is essentially a parasite of the cat, where it lives within the epithelial cells of the small intestines. One way cats become infected is when the oocyst stage is ingested and the infectious tachyzoite stage (motile rapidly dividing form) is liberated from the oocyst in the small intestine (Fig. 11.1). There the tachyzoites will reproduce asexually by binary fission, whereby new tachyzoites will emerge and infect surrounding epithelial cells in the cat intestine. Eventually some of the parasites will produce sexual forms called microgametocytes (male) and macrogametocytes (female) and these initiate the sexual cycle in the feline intestine. Fertilization of female gametes by male gametes results in new oocysts being formed in the gut which are then shed in the cat feces.

© Springer International Publishing AG, part of Springer Nature 2017
B. H. Kwa, *The Parasite Chronicles*, https://doi.org/10.1007/978-3-319-74923-5_11

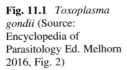

Fig. 11.1 *Toxoplasma gondii* (Source: Encyclopedia of Parasitology Ed. Melhorn 2016, Fig. 2)

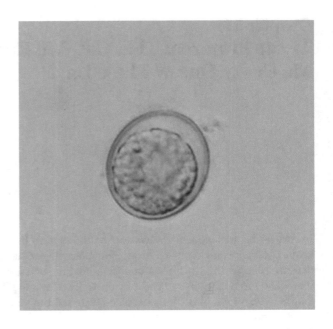

The oocysts when ingested by other animals such as rodents will liberate tachyzoites to infect their tissues. Other incidental hosts include pigs, horses, sheep, goats, hares and numerous other mammals including humans (even birds can be infected). In these hosts, the rapidly dividing tachyzoites would spread and even infect cells outside the intestines such as muscle and neural tissue including the brain (Fig. 11.2). After an initial acute phase when the tachyzoites would divide rapidly, they would eventually slow down and encyst in the tissues, becoming the slowly growing bradyzoite stage within cysts. It is thought that this transition from the tachyzoite form into the encysted bradyzoite form occurs as a protective response by the parasite to host immunity. It is known, for example, that when host immunity is experimentally suppressed in laboratory rat infections, bradyzoites can revert back into tachyzoites and cause a resurgent acute infection. Under normal circumstances bradyzoite cysts would remain in the tissues of the hosts without causing any ill effects.

Other than by ingesting oocysts from items contaminated by infected cat feces, cats may also be infected when they ingest encysted bradyzoites in uncooked infected meat products or from rodent prey. This is a common occurrence that maintains the cycle and it is worth noting that only cats, both the domestic variety as well as wild felines such as bobcats, ocelots, etc., are the only hosts that produce the oocyst stage. However when other animals ingest uncooked meat, such as rodents eating food scraps of raw pork or lamb with bradyzoite cysts, or when pigs are fed raw contaminated meat products, infections may be maintained via the encysted bradyzoites in the food chain.

Fig. 11.2 *Toxoplasma gondii* (Source: Encyclopedia of Parasitology Ed. Melhorn 2016, Fig. 3)

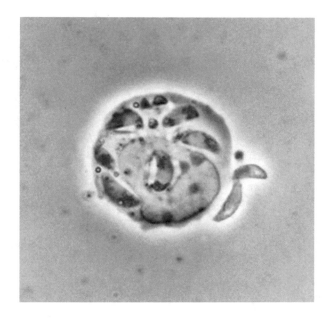

Humans may become incidental hosts through similar routes from both the oocyst stage in cats and from the encysted bradyzoite stage in animal flesh. Thus potentially we may be infected by accidental ingestion of *Toxoplasma* oocysts on contaminated hands as a result of handling the litter box of our domestic pet, especially if it is an outdoor cat which hunts and become infected by consuming infected rodent or bird prey outside the home. Although possible, this route is considered a relatively low risk since infected cats will normally shed oocysts for only a short duration of 2 or 3 days. However humans can also be infected via other routes such as from ingesting contaminated vegetables, fruits, or soil contaminated by oocysts from wild felines. An alternative route would be to become infected by ingesting bradyzoites in insufficiently cooked sausages or rare meat for example (Fig. 11.3).

Most infections in immunocompetent human hosts are asymptomatic but pregnant women, who become infected with an acute infection during pregnancy, are at great risk that their fetus would be seriously affected *in utero*. The rapidly dividing tachyzoites are disseminated in the mother during the acute phase and would infect the developing fetus. Infection during the early part of pregnancy is more dangerous and poses greater risk for abnormalities in the developing fetus. Serious effects include neurological involvement such as intracranial calcifications and hydrocephalus, and may result in death of the fetus. Exposure to the parasite during late pregnancy such as in the third trimester although less severe, may still involve serious ocular pathology such as retinochoroiditis in the infants. It should be noted however that women infected in the past before becoming pregnant are not at similar risk to their fetus. Current drugs for the treatment of toxoplasmosis include pyrimethamine (Daraprim) and sulfadiazine in the United States. Spiramycin is also

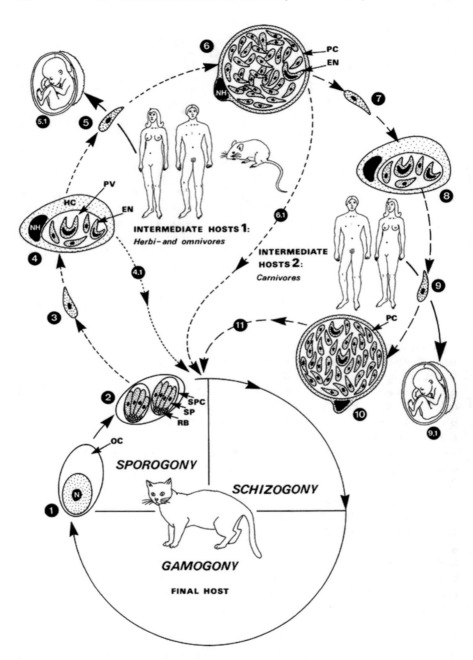

Fig. 11.3 *Toxoplasma gondii* (Source: Encyclopedia of Parasitology Ed. Melhorn 2016, Fig. 1)

routinely prescribed in Europe but is not currently approved in the US, and would require special approval by the FDA.

An article in The Atlantic magazine in 2012 called "How Your Cat Is Making You Crazy" by Kathleen McAuliffe, which was later expanded into a book titled "This Is Your Brain on Parasites" by the same author in 2016, described how *Toxoplasma gondii*-infected rats not only lost their normal fear of the presence of cats and avoiding places which smelled of cat urine, but actually became attracted by the smell of cat urine. The infected rats also became hyperactive and displayed more reckless behavior, which led these rats to be more likely caught and eaten by cats which tend to be attracted by fast moving objects. This will result in a greater probability of the parasite completing its cycle and ensuring species survival of the parasite.

However the book "This Is Your Brain on Parasites" went on further to discuss controversial research that suggested that similar effects may be happening to the behavior of people infected by *T. gondii*. This suggested link to human behavior seems plausible because *T. gondii* will infect host neural tissues including the human brain. For example McAuliffe cited studies suggesting that infected persons may display "personality changes, increased impulsivity, less fear or poor judgment in potentially dangerous situations such as 'when to speed up and pass a car'", implying that *T. gondii*-infected people might be indulging in more reckless driving habits (page 75, ibid.). She also cited research that reported that "people who are profoundly depressed or suicidal are more likely to have brains that show signs of inflammation—an immune reaction to infection" (page 80, ibid.). So are humans similarly showing signs of being manipulated by *T. gondii*? Even though in this case the parasite had inadvertently infected the wrong host, resulting in no evolutionary advantage to the parasite, since infected humans are unlikely to be consumed by cats—we are essentially "dead end" hosts for the parasite. Interestingly, McAuliffe's book further cited reports that "people with schizophrenia are 2 or 3 times more likely to have antibodies against the parasite than those who don't have the disorder" (page 79, ibid.); and that "those who had elevated antibody levels against the parasite were 1.5 times more likely to attempt suicide" (page 80, ibid.).

These suggestions that human behavior is being manipulated by *Toxoplasma gondii* unsurprisingly caused a huge sensation when it came out, especially when it was linked to a familiar household pet. The theory is indeed tantalizing. The evidence is however still sparse and many in the scientific community are not completely convinced. One long-term study, which looked at blood samples from 1000 people born between 1972–1973 in Dunedin, New Zealand for example, could not find any association between the presence of antibodies to *T. gondii* (28% among those who donated blood were positive) and rates of schizophrenia or depression. The researchers were also unable to observe any link of infection to specific behavioral indicators such as criminal convictions, driving offences and accident claims on insurance, which might have suggested altered personality traits or modified behavior. Thus more studies need to be done before we can be sure if human behavior can indeed be manipulated by this parasite. But it is definitely a legitimate question in the study of human parasitology, and this hypothesis that parasites may have the ability to control human behavior is certainly one of the more intriguing conundrums that have emerged in recent years in parasitology.

***

Chagas' disease is another disease caused by a tissue protozoan parasite. It was named after the eminent Brazilian scientist Carlos Chagas who first described the infection and the etiologic agent is *Trypanosoma cruzi* which is named for a fellow Brazilian scientist Oswaldo Cruz, thus giving this parasite an unmistakable Brazilian connection. Chagas' disease is also called American trypanosomiasis as it is found mainly in rural areas of Mexico, Central and South America where an estimated 8 million people are infected. The vector is the triatomine bug which thrives under poor housing conditions such as in crevices of mud walls and in the spaces within thatch roofs commonly found in rural communities of Latin America. Occasionally a few cases have been reported in the southern United States among immigrants or in autochthonous infections (completely indigenous infections and not a result of contracting it in a foreign country) among indigents sleeping in the open where the cycle may be maintained in dogs, opossums and raccoons. Bloodborne infections acquired from blood transfusion or from organ donation had been reported but are now very rare in the US because of the stricter screening of the blood supply that have been introduced.

It was suggested that Charles Darwin had very likely contracted the disease during the time when he was exploring the region near Mendoza east of the Andes in South America, and if correct, its long term consequences could possibly have been what killed him eventually. He had recorded in his diaries of the "Voyage of the Beagle" that he was bitten by the bug *Triatoma infestans* which is the common insect vector for the disease in that region. He noted in his journal for 26 March 1835: "At night I experienced an attack, and it deserves no less a name, of the Benchuca, the great black bug of the Pampas. It is most disgusting to feel soft wingless insects, about an inch long, crawling over one's body; before sucking they are quite thin, but afterwards round and bloated with blood, and in this state they are easily squashed." Some of the chronic illnesses he suffered in his later years, such as his gastro-intestinal symptoms, were thought to have involved the autonomic nervous system and these were consistent with Chagas' disease. Furthermore, Darwin eventually died of heart disease and cardiac failure which could have been indicative of damage to the heart muscles typical of *T. cruzi* infection.

The triatomine bug which is the vector for Chagas' disease is sometimes called the "kissing bug" because it tends to bite around the mouth and eyes of people and suck their blood. However the parasite enters the body only when the bug defecates near the bite and when we rub the feces either into the bite wound or into mucous membranes such as in the eyes or mouth. My students comment that these bugs add the ultimate insult to injury by "kissing, biting and then shitting on us" before leaving! The infective stage that enters the human body is known as the metacyclic trypomastigote, which is found in the feces of the triatomine bug.

These motile flagellated organisms will enter the cells of infected tissues, lose their flagellum and live intracellularly as amastigotes. The amastigotes will occasionally transform back into trypomastigotes and reenter the blood stream where new generations of trypanomastigotes will reinfect other tissues and live within the cells again as amastigotes. Trypomastigotes use the flagellum for motility and can be observed under the microscope to swim rapidly among the red blood cells. When these trypomastigotes in human blood are ingested by another triatomine bug when the person is bitten, the trypomastigotes would transform into an epimastigote stage in the midgut of the bug, and eventually migrate into their hind gut where they develop into the infectious stage known as metacyclic trypomastigote. By this rather complicated life cycle the parasite can now re-enter another human host when the bug defecates and when the metacyclic trypomastigotes are rubbed into our eyes or mouth (metacyclic here refers to the feature that this is the infective stage in the insect vector which would complete its life cycle in the human host).

The names of the various life cycle stages such as trypomastigote, epimastigote, and amastigote refer to the position of a body called a kinetoplast relative to the nucleus of this unicellular parasite (Fig. 11.4). The kinetoplast forms the base from which the flagellum attaches itself to the body of the parasite, and is where the mitochondria are located. The parasite swims forward in the direction of the motile flagellum, exactly opposite from the movement of the mammalian spermatozoa for instance. Thus a trypomastigote has its flagellum emerging from behind the nucleus (posterior station), the epimastigote has its flagellum emerge from mid-body in front (anterior station) of the nucleus, and amastigotes do not have a flagellum. These names were not invented to confuse or complicate the study of these parasites, even though college students suspect that their professors made them up to use as trick questions in parasitology exams! The way these structures appear under the microscope is a useful way to distinguish various life cycle stages as well as the species of the parasite. For example there is one other form—the promastigote—whose flagellum arises from the anterior tip of the parasite, which is not found in *T. cruzi* but is present in the various *Leishmania* species discussed later.

Initially during the acute phase of human infection by *T. cruzi*, the person may experience fever and there is usually swelling and inflammation around where the parasite entered the skin or mucous membrane. A common clinical sign is a distinctive swelling around the eyelid called Romaña's sign on the site where the infective stages are rubbed in. After this acute phase, most infected individuals will then enter a prolonged "chronic indeterminate" phase that may last for years where the person will not show any signs or symptoms of disease, and where no blood stages are ever seen. For some, they will remain asymptomatic for life without any further signs of disease.

However serious complications occur in about 20–30% of cases who will develop debilitating and sometimes life-threatening medical problems over the course of their lives. Charles Darwin may have belonged to this unfortunate category of people who will develop an enlarged and dilated heart that does not pump blood efficiently and have abnormal heart rhythm that may sometimes cause sudden death. The amastigote stage will infect and develop inside the cells of muscle fibers including cardiac muscles, and this would cause serious cardiomyopathy where inflammatory infiltrates and target heart cell lyses are common microscopic findings. Many may

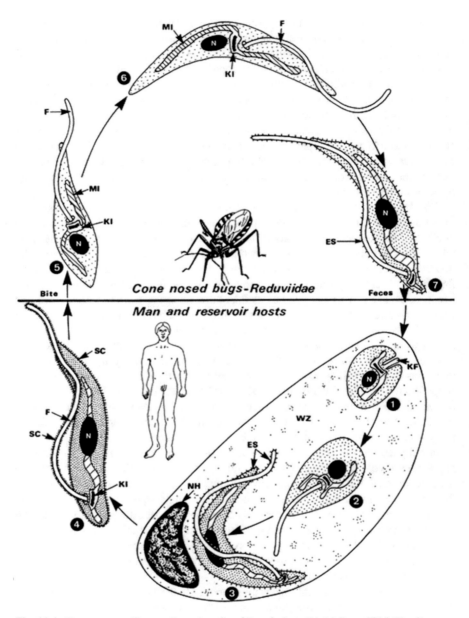

Fig. 11.4 *Trypanosoma* (Source: Encyclopedia of Parasitology Ed. Melhorn 2016, Fig. 2)

also develop necrotic lesions in the autonomic nervous system resulting in gastro-intestinal complications. Amastigotes appear to destroy the parasympathetic neurons which ultimately result in dilated esophagus or colon, leading to severe gastrointestinal problems including difficulty with eating or passing stool.

Furthermore, in people whose immune systems become compromised due to AIDS or chemotherapy for example, Chagas' disease can be reactivated with parasites found in the circulating blood, resulting in fatal disease. Treatment can be successful with benznidazole and nifurtimox but both of them cause serious side effects and should be prescribed under strict medical observation. Public health measures including insecticide spraying of the inside of houses where the triatomine bugs are found, and improvements in the quality of rural housing are the most effective long term solutions.

***

A closely related parasitic disease to Chagas' disease is leishmaniasis, which has a wider geographical distribution that includes parts of Asia, the Middle East, Africa, southern Europe, as well as Mexico, Central America, and South America. As many as 90 countries have reported the disease and it has been estimated that there are at least 1.3 million cases of cutaneous, mucocutaneous and visceral forms of the disease worldwide. Leishmaniasis is also transmitted by an insect vector, in this case the sandfly (Fig. 11.5).

The most common and widely distributed form of leishmaniasis is the cutaneous form and it occurs in the Middle East and North Africa, Central Asia, South Asia and South America. The parasites belong to the following species: *Leishmania tropica*, *L. major*, and *L. aethiopica*, Occasionally *L. donovani* and *L. infantum* may also cause the cutaneous form of disease. Over two thirds of new cutaneous cases reported by the WHO in 2017 occur in six countries: Afghanistan, Algeria, Brazil, Colombia, Iran and Syria. An estimated 0.6 million to 1 million new cases occur worldwide annually. Another form called mucosal leishmaniasis leads to partial or total destruction of mucous membranes of the nose, mouth and throat and is highly disfiguring (Fig. 11.6). The species responsible are *L. braziliensis, L. panamensis, L. guyanensis* and occasionally *L. amazonensis*. Over 90% of mucosal cases occur in the South American countries of Bolivia, Brazil, and Peru, and also in Ethiopia in Africa. However the third and most serious form is visceral leishmaniasis which is fatal in 95% of cases when left untreated and is caused by the species *L. donovani* and *L. infantum*. Visceral organs such as the spleen and liver are infected and the disease also involves fever, weight loss and anemia. It is called kala-azar and is highly endemic in the Indian sub-continent and East Africa, although the rates have been declining in India, Nepal and Bangladesh. The WHO in 2015 estimated that 50,000–90,000 new cases of visceral leishmaniasis still occur annually but 95% of them occur in only seven countries: Brazil, Ethiopia, India, Kenya, Somalia, South Sudan and Sudan.

Leishmanial parasites belong to more than 20 species including *Leishmania donovani* which is the most important since it causes the visceral form or kala azar. The cycle involves sandflies belonging to the Phlebotominae family and in human infections it starts when the sandfly vector bites an infected person and ingests blood with the amastigote stage of the leishmanial parasite living within macrophages. These amastigotes will be liberated from the ingested parasitized macrophages in the gut of the sandfly where they transform into the promastigote

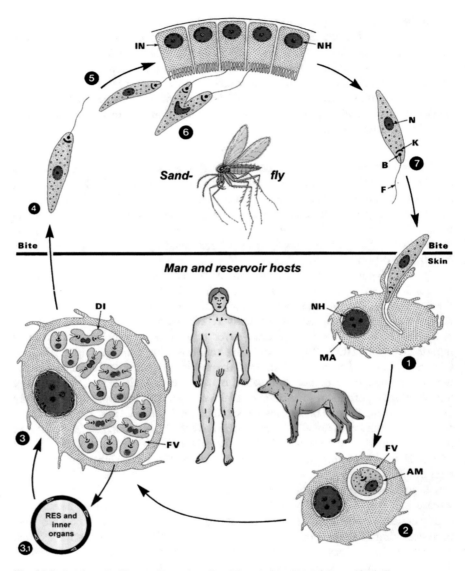

Fig. 11.5 *Leishmania* (Source: Encyclopedia of Parasitology Ed. Melhorn 2016, Fig. 1)

stage. These promastigotes will divide and multiply in the sandfly gut and migrate to the biting mouthpart called the proboscis. When the sandfly bites another person, these infective promastigotes will enter the bloodstream where they will be phagocytized by host blood cells such as macrophages and other mononuclear phagocytic cells. Within the macrophage the promastigotes lose their flagellum and transform into the amastigote stage, and these will infect other cells of various tissues

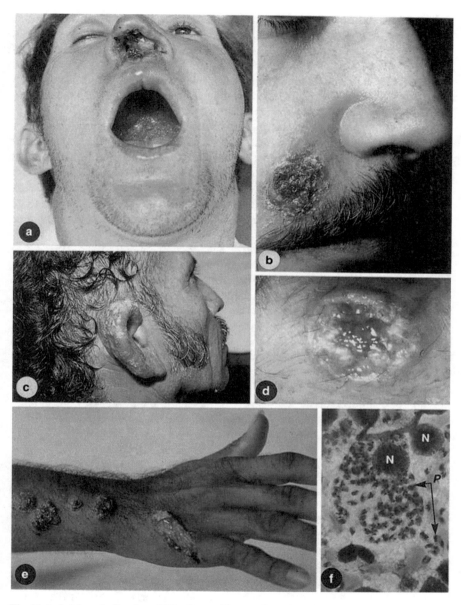

Fig. 11.6 *Leishmania* Species of Skin: Survey [Source: Encyclopedia of Parasitology Ed. Melhorn 2016, Fig. 6 (**a–f**)]

depending on the species of *Leishmania*, and cause the various forms of disease mentioned above. Some will infect other macrophages and the cycle is completed when infected macrophages in the blood are ingested by another sandfly.

The wide clinical presentation seen in cutaneous and mucosal leishmaniasis is mediated by the spectrum of host immunological responses. Mucosal leishmaniasis is associated with an exacerbated T_H1-type response with concurrent high number of CD8+ cytotoxic T cells. These patients exhibit delayed-type hypersensitivity (DTH) and low levels of the regulatory cytokine interleukin-10 (IL-10). These factors result in increased disease severity, where leishmanial parasites migrate to the nasopharyngeal mucosa, multiply in numbers and cause disfiguring lesions.

Patients who present with diffuse cutaneous leishmaniasis also exhibit very severe disease, with extremely high parasite numbers within the disseminated cutaneous lesions. However, this represents the other end of the immunological spectrum, in contrast to the mucosal form of the disease. These individuals exhibit low levels of T_H1 cytokines, have high parasite-specific antibody titers and high levels of IL-10. The category of self-healing cutaneous leishmaniasis appears to be in the middle between the two extremes, reflecting more moderate immune responses overall.

Treatment of leishmaniasis is effective with sodium stibogluconate (Pentostam) and liposomal amphotericin B (AmBisome) which are available in their injectable forms in the U.S. The oral agent Miltefosine was approved by the FDA in 2014 but is contraindicated for pregnant women. Drugs used in treatment of leishmaniasis in some other countries include amphotericin B deoxycholate, pentamidine isethionate, and paromomycin, as well as orally administered "azoles" (ketoconazole, itraconazole, and fluconazole) but the FDA does not currently approve their use for treatment of leishmaniasis. Public health measures are fairly limited since the elimination of sandflies is not practicable because their breeding habitats are widespread in almost any environment with soil and organic matter. Apart from promoting the use of insecticide treated bednets and spraying the insides of houses, other measures include minimizing nocturnal outdoor activities, applying insect repellents on exposed skin and wearing protective clothing.

*** *** ***

Compared to the other tissue protozoan parasites discussed above, *Trichomonas vaginalis* is relatively benign, not life threatening nor causing serious long term medical complications to the patient. It causes an infection called trichomoniasis, sometimes referred to as "trich", and is considered the most common treatable sexually transmitted disease (STD) by the CDC. Although trichomoniasis is more common among women where it infects the vulva, vagina, cervix, and urethra, it can infect men as well, commonly in the male urethra (Fig. 11.7). Symptoms are mild and in women they involve itching, burning, redness or soreness of the genitals, discomfort with urination, and occasional vaginal discharge. In men it may cause itching or irritation inside the penis, burning sensation after urination or ejaculation and more rarely, a discharge from the penis.

More serious complications, as a result of genital inflammation and compromised mucosal integrity, may include increased susceptibility to other STDs including HIV infection. Among pregnant women trichomoniasis increases the risk of preterm

Fig. 11.7 *Trichomonas vaginalis* (Source: Encyclopedia of Parasitology Ed. Melhorn 2016, Fig. 3)

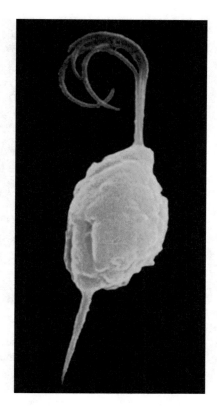

delivery and having low birth weight babies of less than 5.5 lbs. Treatment with either metronidazole or tinidazole is effective and successful.

***

Every year when Memorial Weekend rolls around in Florida and the summer season begins, the Florida Department of Health will launch its "Healthy and Safe Swimming Week" campaign with a series of health advisories to the public. As this is Florida, among the various items is usually a warning about a free living ameba *Naegleria fowleri* that causes a rare but fatal infection of the brain called primary amebic meningoencephalitis (PAM). Local newspapers with an eye to increase sales of their paper tend to call it the "brain eating ameba", which for once happens to be a fairly accurate description and not hyperbole.

Naegleria fowleri is a rather common free living ameba that lives among the detritus at the bottom of most water bodies such as lakes, ponds, canals and slow moving streams, feeding on bacteria among the decaying vegetation (Fig. 11.8). Although they commonly appear in the ameboid trophozoite form at the bottom of ponds, they may occasionally also appear as an actively motile flagellated stage when there is plentiful water. *N. fowleri* amebae are very adaptable to environmental changes and they may transform into a cyst form to survive during dry conditions.

Fig. 11.8 *Naegleriasis* (Source: Encyclopedia of Parasitology Ed. Melhorn 2016, Fig. 1)

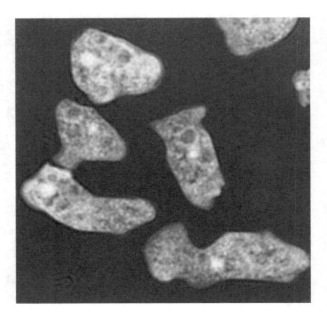

However they generally prefer warmer temperatures and will start to multiply under favorable tropical conditions in summer when there is a lot of rain and surface water temperatures in southern states such as Florida can reach the mid to high 80 °F.

Swallowing *N. fowleri* is not a risk for infection as it cannot survive the stomach acids. However if the ameba is inadvertently sucked up the nose, which can happen in vigorous water activities such as diving or taking a tumble when water skiing in a lake, it can attach itself to the mucous membranes in the nasal cavity and will burrow into the olfactory nerve. It will then follow the route of this nerve and travel directly into the brain. Once it reaches the olfactory bulbs within the brain they will start multiplying and feeding on the brain cells, starting a fulminant infection. Although very rare, with only 34 cases reported in the U.S. during the past 10 years, infection with *N. fowleri* is almost invariably fatal and kills more than 97% of its victims within days. Death usually ensues within a few days due both to brain inflammation as well as rapid and extensive destruction of brain tissue by the rapidly proliferating amebae. There appears to be evidence to suggest that incidence of this infection may be increasing in the U.S. Prior to 2010 more than half of cases came from Florida, Texas and other southern states, whereas recent infections have been reported as far north as Minnesota.

Survival of patients infected with *N. fowleri* is very rare because of the rapid progression and destructiveness of the lesions, giving very little time for medications to work. In the few cases where the patients were successfully treated with Miltefosine, it appears that early and accurate diagnosis and the immediate initiation of treatment might have played crucial roles in saving the victims.

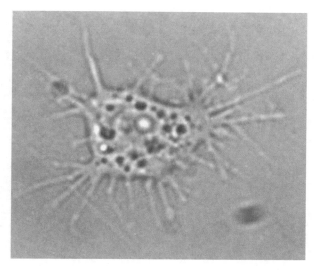

Fig. 11.9 *Acanthamoebiasis* (Source: Encyclopedia of Parasitology Ed. Melhorn 2016, Fig. 2)

Several other free living amebae that belong to the genus *Acanthamoeba* may also infect humans in a variety of ways. As contact lens use has become very common, and proper care and cleansing procedures of lenses not always followed by their users, *Acanthamoeba* species such as *A. castellani* have been implicated to cause *Acanthamoeba* keratitis, which is an infection of the cornea which may lead to severe visual impairment and blindness. Contamination of the lenses and lens cleansing solution from dirty fingers are the most common route of infection. Although topical chlorhexidine and polyhexamethylene biguanide have been used to successfully treat *Acanthamoeba* keratitis, early treatment is important before serious damage to the cornea occurs. Contact lens wearers tend to tolerate minor irritation with lenses and may not seek medical attention until severe symptoms appear, and when treatment becomes more difficult.

Other species of *Acanthamoeba* may enter the skin through a cut or wound and cause a skin infection (Fig. 11.9). More rarely they may proceed to become distributed via the bloodstream throughout the body, causing a disseminated infection in other organs especially the lungs, brain, and spinal cord. One such species is *A. polyphaga* although the taxonomy of many of the free living *Acanthamoeba* is still indeterminate. Although rare, the most serious form of infection by *Acanthamoeba* spp. occurs in the spinal cord and brain causing a disease called Granulomatous Amebic Encephalitis (GAE). A person with GAE may suffer nausea and vomiting, loss of balance, seizures, and hallucinations. Symptoms usually progress over several weeks and often result in death. Successful treatment have however been reported with combinations of pentamidine, sulfadiazine, flucytosine, and either fluconazole or itraconazole.

Chapter 12
Blood Protozoa: In Search of the Holy Grail

The earliest case description of sleeping sickness, or African trypanosomiasis, was reported by the Arab historian Ibn Khaldun (1332–1406) about Sultan Mari Jata, Emperor of Mali. He wrote: "He told me that Jata had been smitten by the sleeping illness, a disease which frequently afflicts the inhabitants of that climate... Those afflicted are virtually never awake or alert. The sickness harms the patient and continues until he perishes". However it was only with the advent of European conquest and colonization of Africa in the nineteenth century that a series of large scale epidemics of sleeping sickness started. The worst epidemic occurred in the Congo Basin around 1889 where it was estimated that between 300,000 and 500,000 people died under the Belgian colonial authorities. This was a consequence of forced dislocation of populations and indentured labor for the purposes of construction of roads and railways, opening of new lands in tsetse infested areas for plantations of cash crops and the collection of rubber, and harsh imposition of new taxes on already vulnerable populations. New and relatively fast means of transport by railways meant too that trypanosome strains were transferred from one area to another, and the "*cordon sanitaire*" maintained in the past by long distances between villages and limited contact between local tribal communities had been broken.

There are two main forms of African trypanosomiasis: *Trypanosoma brucei gambiense* and *Trypanosoma brucei rhodesiense*. The disease caused by *T. b. gambiense* covers 24 countries mainly in west and central Africa and accounts for 97% of all cases. It causes a chronic infection in which some patients may be infected for months or years without symptoms and by the time clinical signs appear, the disease may have progressed to include neurological involvement (Fig. 12.1). *T. b. rhodesiense* on the other hand is an acute disease in which symptoms and clinical signs which involve the central nervous system may appear within a few weeks of infection. It is prevalent mainly in eastern and southern Africa where the WHO reports that currently 13 countries are affected. This form causes a rapid and severe disease but fortunately it represents around only 3% of cases of African trypanosomiasis.

© Springer International Publishing AG, part of Springer Nature 2017
B. H. Kwa, *The Parasite Chronicles*, https://doi.org/10.1007/978-3-319-74923-5_12

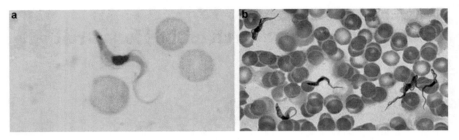

Fig. 12.1 *Trypanosoma brucei* Group of Humans (Source: Encyclopedia of Parasitology Ed. Melhorn 2016, Fig. 3)

Tsetse flies that belong to the genus *Glossina* are the vectors for trypanosomiasis and when an infected fly bites, it will inject the metacyclic trypomastigote (we had discussed similar life cycle stages in the previous chapter) into the skin. This infective stage then finds a lymph vessel to enter, and from there it migrates into the blood circulation. There it becomes the blood stage trypomastigote, reproduces by binary fission and proceed to spread into other body fluids apart from the blood stream, such as lymph and cerebrospinal fluid. Throughout the human body the parasite remains entirely extracellular. When bitten by a tsetse fly, the trypomastigote is ingested into the midgut of the fly where it becomes a procyclic trypomastigote which reproduces by binary fission. It will then migrate out of the midgut towards the anterior of the fly digestive system where it transforms into the epimastigote stage and eventually reaches the salivary glands. In the salivary gland it will develop into the infective metacyclic trypomastigote stage and will enter the human body with the next tsetse bite to complete the cycle (Fig. 12.2).

In the early stages of infection, clinical disease includes headaches, fever, weakness and pain in the joints, lymphadenopathy, and stiffness. An interesting early sign of disease has to do with the appearance of swollen lymph glands along the back of the neck. This was described by an English physician Thomas Winterbottom (1766–1859) and this condition was named Winterbottom's sign. It is an indictment of one of history's most shameful period, that slave traders had used Winterbottom's sign to select or discard the slaves whom they wish to buy and sell. People who are infected with *T. b. gambiense* may not show signs of illness immediately, but after a period of months or even years the parasite would eventually cross the blood-brain barrier and migrate to the central nervous system. There it causes various neurological changes which include sleep disorder (hence the name "sleeping sickness"), sensory disturbances, interference with mobility, ataxia, psychiatric issues, seizures, coma and ultimately death. Infections with *T. b. rhodesiense* will exhibit similar clinical signs and symptoms but the entire process progresses rapidly in a matter of a few weeks or months.

Currently diagnosis is dependent on clinical evaluation for symptoms and signs, and on the Card Agglutination Trypanosomiasis Test (CATT) which was developed in the 1970s for serological screening for *T. b. gambiense* infections. Confirmation is made by microscopic examination of cerebrospinal fluid, performed after lumbar

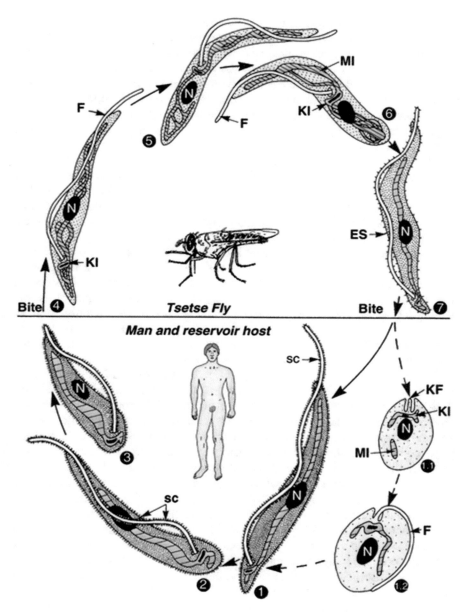

Fig. 12.2 *Trypanosoma brucei* Group of Humans (Source: Encyclopedia of Parasitology Ed. Melhorn 2016, Fig. 1)

puncture and finding the presence of parasites. However since the parasite density is fairly low in *T. b. gambiense* infections, this is an unreliable method and the medical community is still waiting for more rapid and accurate serological techniques. In the case of *T. b. rhodesiense,* since the parasite load in peripheral circulation is higher, blood stages are more easily observed for diagnostic purposes.

Current pharmaceutical treatment of trypanosomiasis is difficult because of the toxicity and complex administration of the drugs currently available. Four drugs have been registered for the treatment of human African trypanosomiasis: pentamidine, suramin, melarsoprol and eflornithine. A fifth drug, nifurtimox, is used in combination under special authorizations. However, all have a certain level of toxicity, and currently pentamidine and suramin are used in the early stages of *T.b. gambiense* and *T.b. rhodesiense* infections respectively.

Eflornithine can be used in monotherapy but only for *T. b. gambiense* infection since it is ineffective against *T. b rhodesiense.* The WHO reports that since 2009, the combination of eflornithine and nifurtimox (NECT) has been adopted as first line treatment for *T. b. gambiense* in endemic countries as this combination minimizes the duration of eflornithine treatment alone and is easier to administer, at the same time making it safer and more effective.

Eflornithine is difficult to administer and requires skilled staff and bulky supplementary material and making the logistics a challenge. To ensure its extensive use by National Sleeping Sickness Control Programs (SSNCP) it has now been distributed free of charge in a single packet containing all the materials, expendables and equipment needed for its administration. WHO has also trained national staff on how to manage the various pharmaceutical drugs used. Melarsoprol is the only treatment available for late stage of *T.b. rhodesiense,* and being also used as second line drug for the advanced stage of *T. b. gambiense* infections.

Vector control measures against the tsetse fly are extremely difficult because the breeding habitats of *Glossina* spp. are scattered over a wide range in the bush and female flies deposit an individual fully-formed larva (viviparous) once every 8–10 days on any piece of shaded soil. The larva will burrow into the soil within minutes, become a pupa underneath the soil and the adult emerge within 3 weeks. Larval and pupal control is thus impossible. Both male and female flies seek blood meals and they bite the human host in the open and during daytime, making bednets not an effective strategy. Furthermore *Glossina* spp. feed on a broad range of animals including all types of wildlife as well as domestic cattle, complicating targeted vector control among specific populations that the flies feed on. Early attempts at habitat control such as clearing the bush are environmentally destructive and beyond the resources of the countries with endemic disease. Thus vector control strategies can target only the adult stages, and such attempts as odor-baited traps, aerial ultra-low volume insecticide spraying and using insecticide-treated live animal baits had only marginal success. Sterile male release techniques had some success in Zanzibar, which is an island, but could not be replicated in the open bush or forested areas on the mainland.

Despite all the challenges, there had been encouraging progress in reducing disease burden, largely due to drug treatment. In 1998 the WHO reported more than 40,000 cases but it was estimated that the true number might have been as high as 300,000. By 2009 after continued control efforts, the number of reported cases dropped below 10,000, and in 2015 only 2804 new cases were reported. However estimates of actual cases put that number closer to 20,000 and with wars, civil disturbances and displaced populations, it is possible that a resurgence of trypanosomiasis may occur at any time.

***

In the history of medicine, malaria seemed to have always been with us, right from the dawn of civilization. Historical writings from independent sources as varied as those written in ancient Sanskrit, Greek, Chinese and Latin have recorded clinical signs and symptoms of malaria with great accuracy and detail from millennia ago. The name itself in English came from the Italian "mal'aria" or "bad air" and that was in turn likely to have been derived from the earlier Latin "malus aria". The ancient Romans avoided living near swampy areas because they associated them with fevers from breathing in the evil vapors. In India, an early Sanskrit medical treatise—the *Susruta*—had described symptoms of malaria, even correctly speculating that it could be attributed to insect bites. Hippocrates in ancient Greece had classified the different types of malaria based on the periodic fevers into quotidian (daily, every 24 h), tertian (every third day, 48 h) and quartan (every fourth day, 72 h) which are still valid today; and by the age of Pericles, references to the disease were quite widespread. The classical Chinese *Nei Jing* ("The Canon of Internal Medicine") had chronicled detailed symptomatology of malaria in 2700 BCE.

Some of the antimalarial drugs currently prescribed today are derived from medicinal plants which have been in use for centuries. Quinine is still an important antimalarial medication today, but it was derived from the Cinchona tree (*Cinchona* spp.) whose use was first observed by European colonialists in seventeenth century Peru (Fig. 12.3). It had been an old remedy for malaria even then, already a part of the pharmacopeia of indigenous peoples in South America for generations. In China, in the second century BCE, the *Artemisia* plants (*Artemisia apiacea* and *A. annua*) were recorded in the ancient Chinese *Materia Medica* ("*Ben Cao*") and in that treatise the two species had been classified as "Qing hao" (*A. apiacea*) and "huang hua hao" (*A. annua*) and were specifically prescribed for treating the periodic fevers seen in malaria. The active ingredient of *Artemisia*, later named artemisinin, was isolated by Chinese scientists in 1971 and derivatives of artemisinin are today among the most potent and effective antimalarial medications available, especially in combination therapies with other drugs. The Nobel Prize in Medicine was awarded to Youyou Tu in 2015 for her research related to the pharmacology of artemisinin and its biomedical properties, thus completing the historical connection of this drug of Chinese origin from the second century BCE to the twenty-first century CE.

Yet despite the accumulated knowledge and wisdom of centuries in the study of malaria and discoveries of medications to treat it, we are not much closer today at eradicating the disease. Why is this so? The answer lies in the superior adaptability that the malarial parasite possesses, compared to anything that humans could devise to destroy it. Its adaptability is in part because the parasite *Plasmodium* has an extremely complex life cycle with different stages and forms which live in different organ systems of its host. Thus pharmaceutical drugs effective against one parasite stage in the red blood cells may be ineffective against another stage within an

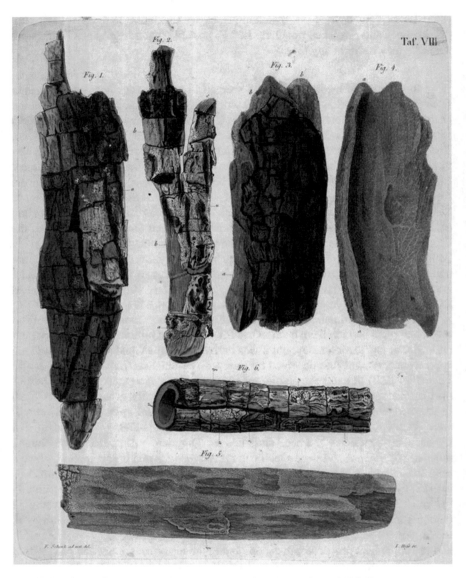

Fig. 12.3 Quinine (Source: Encyclopedia of Parasitology Ed. Melhorn 2016, Fig. 1)

inaccessible location inside the liver. A vaccine against the stage that first enters the blood may stimulate antibodies that kill it, but the same antibodies may not detect a different stage already hidden under the membrane surrounding the cytoplasm of a blood cell. Furthermore the malarial parasite is continually evolving and producing new genetic strains, which become resistant to drugs almost as fast as we can produce new ones. Similarly the mosquito intermediate host will also evolve genetically and produce insecticide-resistant strains within a short time, confounding our attempts to use the same chemical pesticides to eliminate the mosquito vectors for very long. Thus the eradication of malaria is still very much the Holy Grail that many parasitologists have long hoped to solve.

According to the World Health Organization (WHO) as of 2015 there are at least 91 countries which still have on-going malaria transmission, and it was estimated that it puts about 3.0 billion people who live in those areas at risk for infection. In the latest WHO data, released in December 2016, it was estimated that there were 212 million cases of malaria in 2015 that caused 429,000 deaths worldwide. However there is optimism that the overall picture is improving even though the numbers are still staggering. Malaria incidence (rate of new cases) among populations at risk fell by 21% globally between 2010 and 2015; during the same period, malaria mortality rates among populations at risk had decreased by 29%. WHO estimated that 6.8 million malaria deaths had been averted globally since 2001.

Children under five are particularly susceptible and more than two thirds (70%) of all malaria deaths occur in this age group. Nevertheless, even in this category there had been progress made, as the under-5 malaria death rate had fallen by 29% globally from 2010 to 2015 according to the WHO. However in terms of geographical distribution, Africa suffers disproportionately as the region currently has 90% of malaria cases and 92% of the total malaria deaths worldwide. Therefore many of the malaria control initiatives by international organizations are concentrated on the 13 or so countries with the highest prevalence of disease (number of existing cases), mostly in sub-Saharan Africa.

There are four main species of *Plasmodium* parasites that have long been known to cause malaria in humans, and a fifth species more recently discovered in Malaysia which can naturally infect humans even though it is primarily a parasite of macaque monkeys in the wild. The most important of these species is *Plasmodium falciparum* which has the most extensive geographical distribution and may be found worldwide in almost all tropical and subtropical areas. It is also the cause of the highest mortality rates among malaria infections especially in Africa, where it is the predominant species. It multiplies rapidly in the red blood cells where it destroys a large proportion of them and results in severe anemia in the patient. *P. falciparum* also has a tendency to cause blockage of small blood vessels, and when this occurs in the brain, will cause cerebral malaria which is the main cause of death among infected children. The stages in the red blood cells show distinctive characteristics such as the appearance of more than one parasite within a single cell (this is called multiple infection; this should not be confused with mixed infections, which means more than one species of malaria infecting a single person) and unique banana shaped gametocytes which help identification of this species under the microscope. The periodic fever exhibited by this species is tertian (every third day) and in the classical literature it was designated as malignant tertian malaria.

Plasmodium vivax is found in Asia, Oceania, Latin America and parts of Africa, and its prevalence is substantial due to the high population densities in the Asian countries at risk for *P. vivax* infection. People who are negative for Duffy blood group proteins are protected against *P. vivax* infection, and because these populations reside mainly in sub-Saharan Africa, it accounts for the lower prevalence there. This species (as well as *P. ovale*) have stages that remain hidden in the liver which can be reactivated and invade the blood causing relapses of the disease even after months or years. The blood stages seen in thin blood smears under the microscope appear with typical ameboid ring stages, often with characteristic stippling called Schüffner's dots. The periodic fever exhibited by this species is tertian (every third day) and in the classical literature it was designated as benign tertian malaria.

Plasmodium ovale is found mainly in parts of Africa (especially in West Africa) and sporadically in islands of the Western Pacific. Unlike *P. vivax* however, *P. ovale* can infect individuals who are Duffy blood group negative, accounting for its higher prevalence in sub-Saharan Africa. Otherwise *P. vivax* and *P. ovale* are very similar in morphology and are also similar in that both can cause relapses. While Schüffner's dots are often seen in both species, *P. ovale* infected blood cells often appear oval or elongated, hence the name. The periodic fever exhibited by this species is tertian (every third day).

Plasmodium malariae is found worldwide and if untreated, causes a long-lasting, chronic infection that in some cases can last a lifetime. In some chronically infected patients *P. malariae* can cause serious complications such as the nephrotic syndrome. The appearance of "band" forms of blood parasites, where they extend across the entire diameter of the infected red blood cell, is distinctive and may be used to differentiate it from the other species. It is also the only human malaria parasite species that has a quartan fever cycle (every fourth day).

Plasmodium knowlesi is found in many countries in Southeast Asia where it is long known as a natural pathogen of the long-tailed macaque (*Macaca fascicularis*), and pig-tailed macaque (*M. nemestrina*). However in 2004 it was reported for the first time by Balbir Singh and his colleagues in the state of Sarawak in East Malaysia that 208 people had naturally acquired infections of *P. knowlesi* that were confirmed by polymerase chain reaction (PCR). *P. knowlesi* has a 24-h replication cycle and may rapidly progress from an uncomplicated to a severe infection and sometimes death in humans. Morphologically they have features that overlap those of *P. malariae* and *P. falciparum,* making microscopic diagnosis difficult. The periodic fever exhibited by this species is quotidian (every day).

The life cycle of the malarial parasite is very complex and is the bane of students of parasitology (Fig. 12.4). The human host is infected when a female infectious *Anopheles* mosquito inoculates the infective sporozoite stage into the blood stream when it is taking a blood meal (Fig. 12.5). The sporozoite stage remains only for a short period of time (6 h when injected under experimental conditions) in the blood or lymph before it enters the liver cells and matures into a schizont stage which produces the merozoites (this process is called exo-erythrocytic schizogony or the exo-erythrocytic cycle). The merozoites appear in the blood stream usually after 10–15 days, and this period between infection by the sporozoites and the initial appearance of blood stages is called the "pre-patent period". It is sometimes confused with the "incubation period" which in malaria usually refers to the period after

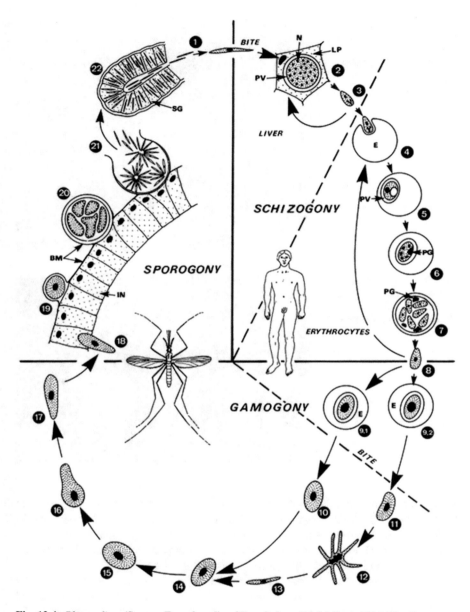

Fig. 12.4 *Plasmodium* (Source: Encyclopedia of Parasitology Ed. Melhorn 2016, Fig. 2)

exposure and before symptoms first appear, which can vary from 7 to 30 days in the different species.

In the case of *Plasmodium vivax* and *P. ovale*, there is a persistent hypnozoite stage that remains dormant but alive in the hepatic cells of the liver for months and

Fig. 12.5 *Plasmodium* Species of Humans (Source: Encyclopedia of Parasitology Ed. Melhorn 2016, Fig. 2)

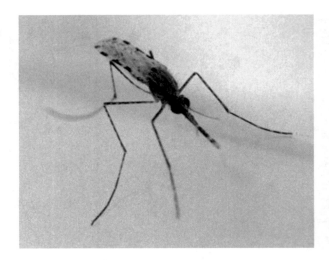

Fig. 12.6 *Plasmodium* Species of Humans (Source: Encyclopedia of Parasitology Ed. Melhorn 2016, Fig. 6)

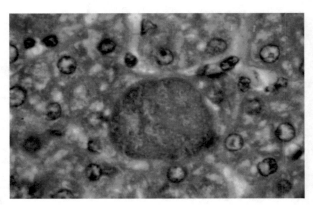

even years, and it is this feature of subsequent re-emerging merozoites from the hypnozoite stage that is responsible for the relapses seen in these two species. In *P. falciparum*, *P. malariae* and *P. knowlesi* however, once the initial merozoites are released into the blood from the liver, the exo-erythrocytic cycle is completed and no parasites remain in the liver, hence no relapses (Fig. 12.6).

The merozoites released from the liver will proceed to infect the red blood cells and initiate the erythrocytic cycle in the blood. The merozoite enters into the cell and starts the development of the trophozoite stage which grows in the cytoplasm of the red cell and feeds on the hemoglobin. The developing trophozoite has the appearance of a signet ring which is why it is sometimes called the ring stage. The trophozoites eventually develop into schizonts which produce numerous merozoites within the red blood cell, and when the infected cell ruptures the merozoites will infect other red cells. This process is called the erythrocytic cycle because it takes place entirely in the red blood cell or erythrocyte. Most of the clinical symptoms of malaria are associated with these blood stages.

When the parasite develops in the red blood cells, metabolites such as hemozoin pigment and other toxic factors accumulate within the erythrocyte. These are released into the bloodstream when the infected cells rupture and invasive merozoites are liberated. The hemozoin and other toxic factors such as glucose phosphate isomerase (GPI) will stimulate macrophages and other cells to produce cytokines and other soluble factors which act to produce fever and rigors and they probably also influence other severe pathophysiology associated with malaria.

During this blood cycle the rupture of infected red cells eventually becomes synchronized, and since the different species have distinctive cycles of 24, 48 and 72 h respectively, they exhibit the typical fever paroxysms at those intervals that physicians use for diagnosis. These blood stages can also be differentiated morphologically under the microscope, since they have the unique features mentioned earlier that distinguish the different species and stages (Fig. 12.7).

After a period of erythrocytic schizogony, some of the trophozoites develop into sexual stages called gametocytes in blood cells. When a mosquito bites the human host and ingests the blood through the proboscis, these male and female gametocytes will enter the mosquito midgut together with the blood meal. This cycle in the mosquito vector is called the sporogonic cycle. The male gametocyte will undergo a process called exflagellation whereby numerous highly motile male gametes are formed. The female gametocyte matures and forms the female gamete, which will be fertilized by one of the male gametes resulting in the formation of a zygote. This all happens within the blood bolus in the mosquito midgut. The zygote then develops into an elongated motile ookinete which invades the wall of the mosquito midgut and develops into an oocyst attached to the gut wall. The oocyst then matures and grows larger, eventually becoming a sporocyst inside which numerous sporozoites develop. When the sporocyst ruptures and releases the sporozoites, they will migrate to the mosquito salivary glands where they remain until the next bite of the mosquito. When the mosquito takes the next blood meal, it will inject salivary fluid through its proboscis that contains anti-coagulants that help in the blood feed. Sporozoites will finally enter the blood with the salivary fluid and perpetuate the malaria cycle, and this cycle is basically similar among all the five species of malaria that infect humans.

However, clinical disease involving *P. falciparum* is particularly serious because red blood cells infected with its trophozoite tend to adhere to the epithelium of blood vessels, and when it involves the capillaries in the brain, will lead to a condition known as cerebral malaria (Fig. 12.8). This is especially dangerous to children under the age of 5 as it is associated with very high mortality rates in this group. Cerebral malaria may also result in impairment of consciousness, seizures, coma, or other neurologic abnormalities.

Apart from cerebral malaria, which is only associated with *P. falciparum*, other severe symptoms associated with *complicated malaria* may occur with all species of malaria. Severe anemia and hemoglobinuria (hemoglobin in the urine) both of which are due to hemolysis (disintegration of red blood cells); abnormalities in blood coagulation; low blood pressure caused by cardiovascular dysfunction; acute

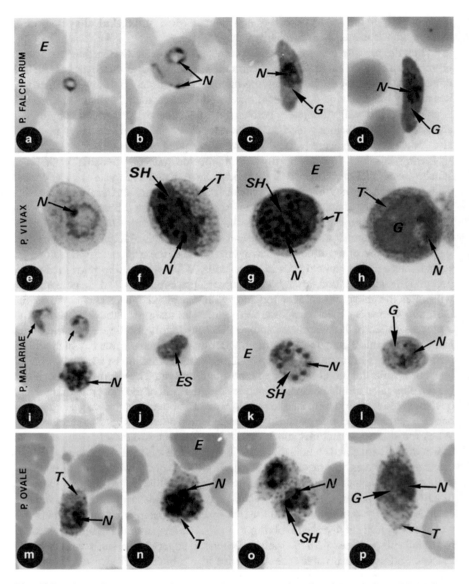

Fig. 12.7 *Plasmodium* Species of Humans (Source: Encyclopedia of Parasitology Ed. Melhorn 2016, Fig. 8)

respiratory distress syndrome (ARDS); acute kidney failure; hypoglycemia (low blood glucose) which in turn can result in metabolic acidosis (excessive acid in the blood and tissue fluid). Hypoglycemia may also occur in pregnant women with uncomplicated malaria, or after treatment with quinine.

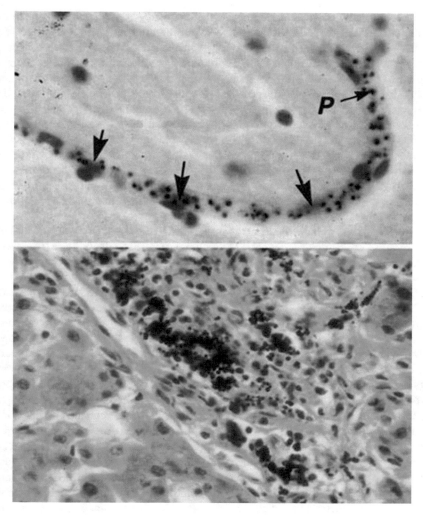

Fig. 12.8 *Plasmodium* Species of Humans (Source: Encyclopedia of Parasitology Ed. Melhorn 2016, Fig. 14)

Uncomplicated malaria, though less severe, is still serious in all species and symptoms may include the classical attack cycle of a cold stage with sensations of chill and intense shivering, followed by a hot stage with fever, headaches, and vomiting; among young children seizures are often featured. Finally, it will be followed by a sweating stage, with a return to normal temperature, and fatigue. Even without the classical attack cycle being presented, uncomplicated malaria could still be a very unpleasant illness associated with fever, chills, headaches, nausea and vomiting, general body aches, exhaustion and malaise.

In countries where malaria is commonplace, residents frequently recognize the symptoms as malaria and treat themselves ("presumptive treatment") with over-the-counter medications. The easy availability of antimalarial drugs without prescription leads to abuse, and is the main cause of drug resistance appearing so rapidly in those regions. I have often seen prescription antimalarial medications such as chloroquine being displayed and sold alongside aspirin and Panadol in pharmacies in some cities and many of the smaller rural towns in Southeast Asia and Latin America. The poor will often buy a few individual pills that they can afford, and stop taking the medication once they feel better. Such widespread use of sub-clinical dosages of antimalarial drugs in the general population results in selection pressure that drives drug resistant strains of the malaria parasite to become predominant. The problem of fake drugs being sold to unsuspecting customers in pharmacies in many parts of the developing world is yet another dilemma, and adds to the other challenges of drug treatment.

Treatment with pharmaceutical agents as listed in the CDC website include chloroquine, atovaquone-proguanil (Malarone), artemether-lumefantrine (Coartem), mefloquine (Lariam), quinine, quinidine, doxycycline (used in combination with quinine), clindamycin (used in combination with quinine), and artesunate (not licensed for use in the United States, but available through the CDC malaria hotline). In addition, primaquine is active against the dormant parasite liver forms (hypnozoites) and prevents relapses. Primaquine should not be taken by pregnant women or by people who are deficient in G6PD (glucose-6-phosphate dehydrogenase). Patients should also not take primaquine until a screening test has excluded G6PD deficiency. Unfortunately there is an increase of resistant malaria strains against these individual drugs, hindering the range of pharmaceuticals we can use in different regions where specific resistance against an individual drug has emerged.

Currently, artemisinin combination therapy (ACT) is recommended by the WHO as a first line treatment of uncomplicated *P. falciparum* malaria. It is recommended for *P. vivax* as well in regions where chloroquine resistance has emerged. Fast acting artemisinin-based compounds such as dihydro-artemisinin, artemether and artesunate are usually paired with unrelated compounds from a different class. Such companion drugs include lumefantrine, mefloquine, amodiaquine, sulfadoxine/pyrimethamine, piperaquine and chlorproguanil/dapsone. These are among the latest in our ever beleaguered armory of pharmaceutical compounds for treatment against malaria, but recent reports that clinical resistance to ACT had appeared in western Cambodia indicate that the struggle between drug resistance and the discovery of new drugs against malaria will be an unending battle that we may have to accept.

Thus, strategies to control malaria would have to depend on a multipronged approach. One basic strategy is to improve accurate diagnosis in resource poor, rural, and underserved communities. Health workers in malaria endemic areas tend to be overloaded with cases and overwhelmed with challenges such as time-consuming laboratory microscopic diagnosis, the lack of working microscopes and lack of well-trained skilled microscopists. Recent rapid diagnostic test (RDT) kits based on immuno-chromatographic methods (a worker in the field only needs to use one drop of patient's blood and note the color change on a strip in a hand held cassette) have been a game changer as they are robust, simple and inexpensive compared to the traditional methods. Once diagnosis is confirmed, the proper medication and dosage could be prescribed once these factors are taken into account: species of malaria, where the infection was acquired and the presence of antimalarial drug resistance in that region, the patient's clinical status such as accompanying illnesses, pregnancy, drug allergies, etc. Improved case management using quick and accurate diagnosis with RDTs alone would be a vast step forward.

The WHO has claimed some success with an initiative called intermittent preventive treatment (IPT). It involves the administration of a curative dose of an antimalarial drug to the *entire population of asymptomatic individuals* at discrete periods of time (frequently more than once). Different individual programs for infants (IPTi), pregnant women (IPTp), children <5 (IPTc) and school children (IPTsc) have been promoted. The program has some notable successes in some countries, but the program is not without controversy however, and critics include those who are concerned that such strategies may increase the drug pressure that in turn increases the likelihood of more drug resistance.

Vector control measures to eradicate mosquitoes have been among the earliest public health strategies attempted against malaria. Mosquito carriers for malaria all belong to the genus *Anopheles*, of which there are in excess of 400 species worldwide, and those that are known to transmit the malaria parasite in nature (vectors) belong to 30–40 species. Among the most important are *Anopheles gambiae* and *An. funestus* in Africa; *An. babirostris* in Asia, and *An. albimanus* in the Americas. The success of species of *Anopheles* mosquitoes as vectors depend not only on their adaptability to survive in different habitats, longevity, flight range and so on, but also whether they are anthropophilic: that is they prefer to seek and feed on human blood. After all, even if a species is very widespread and hardy, but prefers to feed on cattle for instance, they are not likely to be a good vector for human malaria.

Anopheles mosquitoes start their life cycle as eggs laid (ovipositioned) singly by female mosquitoes directly on the water surface. Usually batches of 50–200 eggs are released by a female mosquito per oviposition. These eggs have floats on their sides that prevent them from sinking, and they usually hatch in 2–3 days in the tropical regions where they are usually found. In temperate climates it may take up to 3 weeks for the eggs to hatch.

The larvae that hatch are motile and active, and use a jerky wriggly motion to feed on algae and bacteria just below the water surface. They will periodically return to obtain air by lying parallel to the water surface, since *Anopheles* mosquitoes do not have a special breathing siphon like other types of mosquitoes, they have to position their breathing spiracles on their abdomen next to the air-water interface. Thus entomologists and mosquito control personnel can quickly identify *Anopheles* mosquitoes just by observing their larvae lying parallel to the water surface. The larvae usually live for about a week as they feed and grow in size, and pass through 4 molting stages called instars, when they discard their exoskeleton each time they outgrow them. The larvae are extremely versatile and adaptable and able to survive in habitats as different as fresh water or salt-water marshes, clean fresh water or stagnant drains, mangrove swamps, rice fields, grassy ditches, the edges of streams and rivers, and small temporary rain pools such as those left by tire tracks on unpaved roads. Many breed in open, sun-lit pools while other species are found only in shaded breeding sites under forest canopies. Some species prefer habitats with vegetation while others prefer habitats that have none. Even tree holes and the leaf axils of some plants may serve as habitats for *Anopheles* larvae, although this is less common. This variety of breeding habitats therefore makes conventional larviciding methods like spraying with insecticides against *Anopheles* mosquitoes very challenging.

The next stage that occurs is the pupa which is a dormant stage which does not feed. The pupa is shaped like a comma and clings to the surface of the water to breathe. Although little activity can be seen superficially, within the pupa an extremely vigorous metamorphosis is taking place and a complete adult mosquito, with wings, legs, head with eyes and antennae, is being formed underneath the pupal covering. The entire phase may be completed in 2–3 days before the pupal surface splits and the adult mosquito emerges. The entire process from oviposition to adult mosquito emergence usually takes between 10 and 14 days in the tropics.

Adult mosquitoes mate within a few days after emerging from the pupa, and in most species the males will form large swarms around dusk, and the females fly into the swarms to mate while airborne. Males feed on nectar and usually do not survive for more than a week. Females will also feed on sugar sources for energy, but soon after mating will require a blood meal to obtain protein for the development of the eggs. After obtaining a full blood meal, the female will rest for 2–3 days to digest the blood and develop the eggs. Once they are fully developed, the female will oviposit the eggs and then resumes host seeking. The females mate only once and do not have to mate again as spermatozoa are stored in their spermatheca for subsequent fertilization of the ova.

The cycle repeats itself until the female dies after several egg laying cycles. Female *Anopheles* mosquitoes usually live for 1–2 weeks in nature in tropical regions. Their chances of survival depend on temperature and humidity, and also their ability to successfully obtain blood meals and to find resting places hidden from mosquito predators such as small lizards, frogs and birds.

Vector control methods involve attacking the appropriate stages in the life cycle of the mosquitos. The egg and the process of ovipositioning by the female mosquitoes are stages vulnerable to environmental strategies such as "source reduction", which use methods to remove environmental sites suitable for egg laying. At one end of the scale, massive engineering methods of environmental manipulation to control malaria vector mosquitoes had been successfully pioneered by the Tennessee Valley Authority in the U.S., which depended on altering water levels using a network of levies and dams to control water flow. This enabled the public health engineers to implement widespread ecological disruption of mosquito larval habitats and egg laying sites on a large geographical scale. While spectacularly successful in a rich industrial nation, these methods are beyond the economic resources of poor countries.

Small scale mosquito control measures in endemic countries, while appearing to be an ideal method which is inexpensive, decentralized and low-tech, have not been very effective in practice. The reason is that in developing countries, many of the habitats where the mosquito larvae breed are small, spotty and scattered over large areas which are difficult to reach on the ground. They are also often transient as some places dry up and new areas are flooded which makes it difficult to direct control teams to ever changing new locations. Management of irrigation canals, control of aquatic weeds, draining of swamps, proper maintenance of small rural dams to direct water flows, are all important elements in the eradication of breeding sites and are dependent on efficient and uncorrupt bureaucratic control by effective local authorities.

Similarly, appropriate pesticides used to kill the larvae (larviciding) need to be carefully chosen. There is now less reliance on the use of organophosphate insecticides partly because of environmental concerns and partly because the mosquitos have evolved resistance to many of them. The earlier use of organochlorines have been long stopped because they are persistent in the environment and are damaging to wild life, as they bio-accumulate in the food chain. Thus the selection and management of the appropriate type of environmentally friendly larvicides used are not as simple as they first appear. For example oils sprayed on the water surface to suffocate the mosquito larvae and pupae in breeding sites have to be biodegradable to reduce environmental impact. This means that accurate records of the times and exact location of spraying have to be kept to ensure that the correct locations are re-treated after a specific duration, as the oils would have biodegraded. Rainfall and storm events at the locations have to be taken into account as they wash away the surface oils, and affect the duration when the treatment would have to be re-applied.

Use of biochemical toxins from the bacterium *Bacillus thuringiensis* var. *israelensis* (Bti) is ideal because they are very specific, only targeting mosquitoes, black flies, and midges, and do not pose a risk to fish, mammals, birds and other non-pest insects. However they may have to be applied in pellet form in some areas to ensure that they reach the water where the mosquito larvae live, as aquatic vegetation, over-hanging shrubs, and plant canopies may obstruct liquid spraying. If the depth of the water is too great the pellets sink to the bottom and the toxins will not reach the surface-dwelling larvae.

Insect growth regulators such as methoprene are specific in only affecting mosquitos and are therefore optimal and environmentally friendly larvicides. However they are relatively expensive and tend to be diluted by the frequent tropical rainfall in the countries where malaria is endemic, requiring frequent re-application, making it cost prohibitive.

Furthermore, the broad use of insecticides in the general environment ultimately will result in the appearance of insecticide resistance. Usually this occurs when some areas, for operational reasons such mismanaged insecticide application, will result in sub-minimal coverage of the chemical products in the breeding habitats. Thus inadequate levels of insecticides would select resistant subpopulations of mosquitoes to eventually become the predominant strain.

Fogging with adulticide is not very effective as their action works for only a very short and transient period, when adult mosquitoes are killed by the aerosolized insecticide. They also indiscriminately kill other non-pest insects such as honey bees, butterflies, and dragonflies. Indoor residual spraying (IRS), another method of mosquito control, was at one time thought to be an ideal control method as it is not broadcast into the general environment, but limited only to the insides of houses. They kill *Anopheles* mosquitoes resting on walls and hiding places such as under the bed or under roof rafters which have been sprayed. However they do not directly prevent the mosquitoes from biting the residents, since they only kill the mosquitoes resting after a blood meal. Subsequent studies showed that for IRS to be successful, at least more than 80% of households in a community would have to be sprayed. It is also a labor intensive operation that requires residents to cooperate by allowing the public health personnel to enter their houses and spray their indoor surfaces, and skeptical residents have to be convinced that the spraying is effective since in the short term they would not observe any reduction in mosquito bites.

A much more successful vector control method had been the promotion of insecticide treated nets (ITNs). These bed nets are usually made of polyester, polyethylene, or polypropylene fabrics which have been treated with pyrethroid insecticides such as permethrin. ITNs pose very low health risks to humans and other mammals, and are only toxic to insects and kill them even at very low doses. Since *Anopheles* mosquitoes are generally nocturnal biters and also prefer to bite indoors, they work very well in protecting individuals sleeping under the ITN. ITNs not only act as barriers to protect the people sleeping under them, but since permethrin and related pyrethroids have a repellent quality as well, houses with ITNs have actually fewer mosquitoes entering them. Additionally, it had been shown that when high community coverage with ITNs is achieved, there is a total reduction in mosquito numbers and life span. Therefore the benefits will be conferred on the entire community, regardless of whether or not an individual house is using an ITN. However, to achieve such a community-wide effect, more than half of the households must be using ITNs.

ITN programs are relatively easy to implement, especially in low resource settings. The nets are inexpensive, and they have to be treated only every 6–12 months by simply dipping them in a mixture of water and insecticide and drying them shaded from direct sun. Unfortunately the cost of frequent retreatment may discourage more widespread acceptance of ITNs and recent studies that show the beginnings of permethrin resistance may further undermine the effectiveness of ITNs.

In response, the WHO has been working with several companies to evaluate recently developed long-lasting insecticide-treated nets (LLINs) under the WHO Pesticide Evaluation Scheme (WHOPES). These LLINs use various fabrics such as polyester, polyethylene, or polypropylene with newer pyrethroid derivatives such as deltamethrin and alpha-cypermethrin chemically incorporated in the fibers rather than just coating the surface. Some of these LLINs being evaluated appear to maintain effective levels of insecticide for at least 3 years, even after repeated washing, and should address some of the current challenges with ITNs.

Recent advances in using the CRISPR gene editing techniques to genetically modify mosquito vector strains have also been promising, at least in controlled experiments. Essentially, mosquitoes with the inserted genes interfere with the malarial parasite's ability to migrate to the mosquito salivary glands and complete the cycle. If they can be scaled up and implemented in the general environment, they could be another addition to our armory of methods to control malaria.

The WHO has recently been evaluating its overall strategy in malaria elimination worldwide and for those with an interest in WHO's global efforts to control and eliminate malaria, the Global Technical Strategy for Malaria 2016–2030 (GTS) adopted by the World Health Assembly in May 2015 is essential reading, and can be found on the WHO website.

Finally, malariologists have longed dreamed of the ultimate "magic bullet" to combat malaria—a vaccine. While we seem to be tantalizingly close to having success in finding a malaria vaccine in recent decades, the difficulties are still formidable. As we have discussed, the life cycle stages of the malarial parasite are very complex and heterogeneous, but in addition to that, these stages are capable of exhibiting antigenic variation that evades host immune responses. Despite the challenges, studies using attenuated (artificially weakened) parasite stages in animal models had demonstrated the feasibility of achieving sterile protective immunity, and this has encouraged the on-going search for a human vaccine.

Generally there are three main types of human vaccine candidates: (a) transmission blocking vaccines (TBVs), (b) pre-erythrocytic vaccines and (c) blood stage vaccines. TBVs attempt to block the parasite life cycle in ways that disrupt the transmission of malaria in entire communities. Essentially the idea is to vaccinate people to produce antibodies in their blood that will disrupt the ability of ookinetes and oocysts to develop in the mosquito. These antibodies will be taken up by the mosquito when it blood feeds and will enter the mosquito midgut together with the gametocytes in the blood meal. However the gametocytes would have been damaged by the antibodies and the development of the ookinete and oocyst would also be affected and they will not produce sporozoites normally. Since viable sporozoites will then not be produced in the mosquitoes, this vaccine would gradually reduce new infections in the entire community even if they do not protect individuals from acquiring the disease. The leading TBV candidates are the *P. falciparum* ookinete surface antigens vaccine Pfs25 and Pfs28, and the *P. vivax* equivalents Pfv25 and Pfv28. Pfs25 is expressed as recombinant proteins cross-linked to Exo-Protein A and presented as nano-particles to enhance its immunogenicity. These are in early trials and will hopefully add to our available tools against the disease. However TBVs do not have the advantage of natural immune boosting as people are not exposed to the antigens in nature, unlike the other vaccine candidates.

Pre-erthrocytic vaccines, in the second category, are protective vaccines that kill sporozoites and the liver stages to prevent them from developing into merozoites that emerge into the blood and infect human blood cells. This strategy targets the early stages when the parasite first enters the body and when they are at their most vulnerable. Currently one candidate RTS,S is the furthest along in human trials. It is an injectable vaccine that provided partial protection in young children in Phase 3 clinical trials and it was announced in early 2017 that it will be further assessed in pilot implementation projects in three sub-Saharan African countries: Ghana, Kenya and Malawi, in partnership with the WHO. Another exciting candidate in this category, called PfSPZ vaccines, are radiation-, chemo-, and genetically-attenuated, aseptic, purified, cryopreserved, live whole *P. falciparum* (Pf) sporozoite (SPZ) vaccines. It had been shown to confer >90% protection against controlled human malaria infection in clinical trials in the U.S., Germany, Mali, and Tanzania, and to be protective against intense natural transmission of malaria in Africa. I am very proud to note that one of my first doctoral students who graduated with her PhD from the University of Malaya in the early 1980s when I was teaching there, "Betty" Kim Lee Sim, is a pivotal part of the research team at Sanaria Inc. ("san'aria" is Italian for "healthy air", a witty antonym of the Italian "mal'aria" for "bad air") that developed the "impossible to many" manufacturing process for PfSPZ. Among all the bright and highly successful doctoral students that I had guided over the years, Kim Lee stands out as the most brilliant among them.

Kim Lee had won a WHO TDR Scholarship to study for her PhD under the mentorship of Mak Joon Wah and myself in Kuala Lumpur, and had gone on to complete her post-doctoral research with Willy Piessens and Dyanne Wirth at the Harvard School of Public Health. She had later met her malariologist husband Steve Hoffman, who was then a dashing Indiana Jones-like US Navy Medical officer based in Jakarta, Indonesia. Both of them had "thrown caution to the wind" and had started biotech companies—Kim Lee with Protein Potential LLC and Steve with Sanaria Inc. This was a high risk venture into the wild and wooly west of venture capital, angel investors, scientific review boards and partners from big pharmas, quite a ways removed from the ivory tower of academia. However it is this sense of adventure that had always driven those who worked with parasites and tropical diseases, and for those young and newly minted doctors entering this world and looking for models to follow, I would advise you to hitch your wagon to these stars.

The third category of vaccine candidates is the blood-stage vaccines. These vaccines do not attempt to prevent infection but rather to prevent the development of disease. The idea is that if the vaccine hampers the erythrocytic cycle from occurring, the clinical symptoms, which are mostly a result of destruction of blood cells, would be eliminated or reduced. Encouraging trials in Mali and Thailand may provide us with one more vaccine that could be used together with the vaccines targeting other stages.

These recent advances in vaccine development are very encouraging, as even a partially efficacious vaccine would add to our ability to eliminate malaria: as one more tool in a comprehensive toolkit that will complement others such as LLIN bed nets and pharmaceutical treatment like ACT. Malaria is perhaps the greatest challenge in our endeavor to eliminate parasitic disease among human populations and we are now perhaps closer than ever before in this quest. Hopefully the Holy Grail of parasitology will ultimately be found within the lifetime of my grandchildren.

Epilogue

General Travel Precautions

While this book discusses what appears to be a huge inventory of parasitic diseases, they should not inhibit the traveler from venturing out into the world since a few common sense precautions would be enough to avoid the vast majority of the parasites mentioned. The author is an avid foodie and have been eating street food from street vendors around the world for decades with an enthusiasm that few of the readers would have matched, and yet have avoided any infection (Fig. A1).

The first rule is to never eat anything raw or uncooked. Therefore you should avoid salads, cut fruits, fruit juices, popsicles and the like, and avoid ice with drinks such as sodas since you should always suspect that the ice could have possibly been made with contaminated water. Follow these rules of avoiding uncooked food even in upscale hotel restaurants. Do not be lulled into a false sense of security just because you are in a fancy air conditioned environment where there are clean table cloths and folded napkins and where you are served by impeccably dressed waiters. What happens behind closed doors inside the kitchen has a greater impact on your health and well-being than the splendid surroundings of the restaurant itself. Cooked food are generally safe as most parasites do not survive normal cooking temperatures. So if they are braised, boiled, roasted, grilled, broiled, baked, stir fried or deep fried they are quite safe. I personally avoid eating raw fish such as those in sashimi and cerviche, as noted in the previous chapters, but that is a judgement call as most people have decided that the risks are small enough to justify ignoring my advice.

Whether I eat from street vendors in Ayutthaya or from the little open tea stalls in Pune, or the night market in the Muslim quarter in Xi'an or at a coffee stand in Tegucigalpa, I follow the same rules. First I look for a busy and obviously popular vendor with many locals waiting in line—that is a sure sign that the food is good and the local people trust the hygiene of that place. Show a friendly smile and watch the cooking process with interest; use hand signs and body language to ask about what they are cooking even when you don't know the local lingo. All the while check out

© Springer International Publishing AG, part of Springer Nature 2017   169
B. H. Kwa, *The Parasite Chronicles*, https://doi.org/10.1007/978-3-319-74923-5

Fig. A1 Author sampling street food, Hanoi 1997. From collection of author

the general cleanliness of the eating utensils, plates, cups etc. Nowadays most places around the world use disposable plastic and polystyrene ware, and while not environmentally friendly, are generally hygienic. If you see that what you are about to eat is cooked in front of you and the other customers, you can be quite assured that the general level of hygiene and cleanliness is fine as the locals wouldn't be returning to that stall if it was otherwise. If the food is not being freshly cooked but looked as if it had been displayed and sitting on a tray for some time, I would give a friendly smile and move on.

As for drinks, I would stick to hot coffee or tea (which necessitate boiling the water), or if they have cold sodas in a can or bottle, I will tell them "no ice" by hand signs and drink straight from the can or bottle. Bottled water is ubiquitous and incidentally beer is one of the safest things to drink. I try to eat a lot of fruits whenever I travel and will buy uncut fruits from street vendors and market places. Local fruits are usually cheap, nutritious and safe to eat when you cut or peel them yourself. Discovering new kinds of tropical fruits that you have never eaten before is one of the joys of travel. Pack a small pocket knife in your check-in luggage and you will find that it is a very useful companion to have on a train or bus.

For specific travel advisories, including vaccines and anti-malarial medications, check out the following excellent websites, as public health bulletins change frequently depending on the current situation in different regions of the world and where new disease outbreaks may have recently occurred. The following websites are updated frequently and are the most reliable sources for public health advisories regarding the countries or regions to which you are planning to travel.

The Center for Disease Control and Prevention (CDC) has a continuously updated website which gives you comprehensive and detailed information for all travel related advice.

This should be your first source for travel advisories: https://www.nc.cdc.gov/travel/

They also have a webpage just for vaccines: https://www.cdc.gov/features/vaccines-travel/index.html and another for malaria: https://www.cdc.gov/malaria/travelers/ and updates of recommended anti-malarial medications: https://www.cdc.gov/malaria/travelers/drugs.html There is even a chapter regarding mosquitoes and biting arthropods: https://wwwnc.cdc.gov/travel/yellowbook/2018/the-pre-travel-consultation/protection-against-mosquitoes-ticks-other-arthropods

Another excellent source of information and advice is that provided by the World Health Organization (WHO): http://www.who.int/ith/en/ The WHO organizes the information into categories such as current disease distribution maps, vaccines, and general precautions, and I would highly recommend that you visit their webpage.

Further Reading

Chapter One

Chandler AC, Read CP (1955a) Introduction to parasitology. Wiley, New York & London

Desowitz RS (1981) New Guinea Tapeworms & Jewish Grandmothers. Avon Books, New York

Faust EC, Russell PF (1964) Craig and Faust's clinical parasitology. Lea & Febiger, Philadelphia

Kwa BH (1972a) Studies on the sparganum of *Spirometra erinacei*. I. The histology and histo-chemistry of the scolex. Int J Parasitol 2:23–28

Kwa BH (1972b) Studies on the sparganum of *Spirometra erinacei*. III. The fine structure of the tegument in the scolex. Int J Parasitol 2:35–43

Smyth JD (1962) Introduction to animal parasitology. The English Universities Press, London

Smyth JD (1969) The physiology of Cestodes. Oliver & Boyd, Edinburgh

Chapter Two

Kreston R(2012) Man's best friend, the Turkana tribe & a gruesome parasite. http://blogs.discovermagazine.com/bodyhorrors/2012/01/26/463/#.WZwbvFGQwUc

Markell EK, Voge M (1981) Medical parasitology. Saunders, Philadelphia

Wu YL (1995) Memories of Wu Lien-Teh: Plague fighter. World Scientific Publishing, Singapore

Zhang T, Zhao W, Yang D et al (2015) Human cystic echinococcosis in Heilongjiang Province, China: a retrospective study. BMC Gastroenterol 15:29–34

Chapter Three

Cox FEG (2002) History of human parasitology. Clin Microbiol Rev 15:595–612

Del Brutto OH, Garcia HH (2015) *Taenia solium* cysticercosis – the lessons of history. J Neurol Sci 359:392–395

Flisser A (2013) Epidemiology of neurocysticercosis in Mexico: from a public health problem to its control. In: Sibat HF (ed) Novel aspects on cysticercosis and neurocysticercosis.

https://www.intechopen.com/books/novel-aspects-on-cysticercosis-and-neurocysticercosis/epi demiology-of-neurocysticercosis-in-mexico-from-a-public-health-problem-to-its-control

Galan-Puchades MT, Fuentes MV (2013) *Taenia asiatica*: the most neglected human Taenia and the possibility of cysticercosis. Korean J Parasitol 51:51–54

Kwa BH, Liew FY (1977) Immunity in Taeniasis—Cysticercosis. I. Vaccination against *Taenia taeniaeformis* in rats using purified antigen. J Exp Med 146:118–132

McLachlan RS (2010) Julius Caesar's late onset epilepsy: a case of historic proportions. Can J Neurol Sci 37:557–561

Quackwriter (2015) 'Eat! Eat! Eat!" Those notorious tapeworm diet pills. http://thequackdoctor. com/index.php/eat-eat-eat-those-notorious-tapeworm-diet-pills/

Chapter Four

Daley J (2016) Ancient "poop sticks" offer clues to the spread of disease along the Silk Road. http:// www.smithsonianmag.com/smart-news/ancient-poop-sticks-offer-clues-spread-disease-along- silk-road-180959900/

Davendra C (1991) Integrated animal-fish-mixed cropping systems. http://www.fao.org/docrep/ 004/ac155E/AC155E10.htm

Yeh H, Mao R et al (2016) Early evidence for travel with infectious diseases along the Silk Road: intestinal parasites from 2000 year-old personal hygiene sticks in a latrine at Xuanquanzhi relay station in China. J Archeol Sci Rep 9:758–764

Chapter Five

Araujo A, Ferreira LF (1997) Paleoparasitology of schistosomiasis. Mem Inst Oswaldo Cruz 92:717

Berry-Caban CS (2007) Return of the god of plague: Schistosomiasis in China. J Rural Trop Pub Health 6:45–53

Burke ML, Jones MK, Gobert GN et al (2009) Immunopathogenesis of human schistosomiasis. Parasite Immunol 31:163–176

Di Constanzo J (2002) Gastrointestinal diseases of Napoleon in Saint Helena: causes of death. Sci Prog 85:359–367

Ferguson MS, Bang FB (2016) Chap VI. Schistosomiasis. In: US Army Med Dept Off Med Hist Vol 5. http://history.amedd.army.mil/booksdocs/wwii/communicablediseasesV5/chapter6.htm Accessed 15 July 2016

Garfield E (1986) Schistosomiasis: the scourge of the Third World. Part 1. Etiology. In Essays. Inform Sci 9:65–69. http://www.garfield.library.upenn.edu/essays/v9p065y1986.pdf

Kloos H, David R (2002) The paleoepidemiology of schistosomiasis in ancient Egypt. Res Hum Ecol 9:14–25

Kreston R (1981) The fluke that thwarted an invasion. In: Discover mag. http://blogs. discovermagazine.com/bodyhorrors/2014/09/30/fluke-china-schistosoma/#.WZ2rqVGQwUc

Kwa BH, Li BW (1994) An unwelcome consequence of China's reforms: the resurgence of schistosomiasis. Development – J Soc Int Dev 3:36–38

Ruffer MA (1921) Studies in the palaeopathology of Egypt. University of Chicago Press, Chicago

Chapter Six

Kwa BH (2008) Environmental change, development and vectorborne disease: Malaysia's experience with filariasis, scrub typhus and dengue. Environ Dev Sustain 10:209–217

Kwa BH, Mak JW (1980) Specific depression of cell mediated immunity in Malayan filariasis. Trans R Soc Trop Med Hyg 74:522–527

Kwa BH, Mak JW (1987) Filarids (excluding D. immitis). Chapter 10. In: Soulsby EJL (ed) Immune responses in parasitic infection: immunology, immunopathology and immunoprophylaxis. CRC, Boca Raton, FL, pp 223–249

Rao UR, Vickery AC, Kwa BH et al (1996) Regulatory cytokines in the lymphatic pathology of athymic mice infected with Brugia malayi. Int J Parasitol 26:561–565

Sim BKL, Kwa BH, Mak JW (1982a) Immune responses in human Brugia malayi infective infections: serum dependent cell-mediated destruction of infective larvae in vitro. Trans Roy Soc Trop Med Hyg 76:362–370

Sim BKL, Mak JW, Kwa BH (1982b) Immunoglobin levels in various clinical groups of human Brugian filariasis in Malaysia. Zeitschr Parasitenk 69:371–375

Sim BKL, Mak JW, Kwa BH (1983) Effects of serum from treated patients on antibody-dependent cell adherence to the infective larvae of Brugia malayi. Am J Trop Med Hyg 32:1002–1012

Sim BKL, Kwa BH, Mak JW (1984) The presence of blocking factors in Brugia malayi microfilaraemic patients. Immunology 52:411–416

Sim BKL, Kwa BH, Mak JW (1987) Human in vitro immune reactions to animal filariids. Trop Med Parasitol 38:11–14

Chapter Seven

Botto C, Basanez M-G, Escalona M et al (2016) Evidence of suppression of onchocerciasis transmission in the Venezuelan Amazonian focus. Parasit Vectors 9:40. https://parasitesandvectors.biomedcentral.com/articles/10.1186/s13071-016-1313-z

D'Heurle D, Kwa BH, Vickery AC (1990) Ophthalmic dirofilariasis. Ann Ophthal 22:273–275

Tamarozzi F, Halliday A, Gentil K et al (2011) Onchocerciasis: the role of Wolbachia bacterial endosymbionts in parasite biology, disease pathogenesis, and treatment. Clin Microbiol Rev 24:459–468

Chapter Eight

Araujo A, Ferreira LF, Confalonieri U et al (1988) Hookworms and the peopling of America. Cad Saude Publica Vol 4. http://www.scielo.br/scielo.php?script=sci_arttext&pid=S0102-311X1988000200006

Chandler AC, Read CP (1955b) Introduction to parasitology. Wiley, New York

Cohen J (2012) Native Americans hailed from Siberian highlands, DNA reveals. In: History. http://www.history.com/news/native-americans-hailed-from-siberian-highlands-dna-reveals

Corrales LF, Izurieta R, Moe CL (2006) Association between intestinal parasitic infections and type of sanitation system in rural El Salvador. Trop Med Int Health 11:1821–1831

Cruz LM, Allanson M, Kwa B et al (2012) Morphological changes of Ascaris spp. eggs during their development outside the host. J Parasitol 98:63–68

Kreston R (2014) Poisoned with parasites. In: Discover mag. http://blogs.discovermagazine.com/bodyhorrors/2014/12/27/parasite-poison-2/#.WZ3C0FGQwUc

Livingston D (2003) The flood and subsequent civilization. http://davelivingston.com/postfloodciv.htm

Reinhard KY (1992) Parasitology as an interpretive tool in archaeology. Am Antiquity 57:231–245

Sorenson JL, Johannessen CL. (2004) Scientific evidence for pre-Columbian transoceanic voyages to and from the Americas. In: Sino Platonic Papers 133. http://www.sino-platonic.org/abstracts/spp133_precolumbian.html

Welsh EC (2012) Parasitic nematodes in humans: exploring the host-parasite dynamic through historical, biological, and public health evaluations of infection. In: Univ Louisville Inst Reposit http://ir.library.louisville.edu/cgi/viewcontent.cgi?article=2545&context=etd

Xu M (2016) Olmecs: Transpacific contacts? In TTZ Library. http://ttzlibrary.yuku.com/topic/475/Olmecs-Transpacific-contacts#.WZ26llGQyyo. Accessed 20 Sept 2016.

Chapter Nine

Ambu S, Kwa BH, Mak JW (1982) Studies on the experimental chemotherapy of *Angiostrongylus malaysiensis* infection in rats with mebendazole and levamisole. Trans R Soc Trop Med Hyg 76:458–462

Ambu S, Kwa BH, Mak JW (1985) Histopathology of rats infected with *Angiostrongylus malaysiensis* before and after drug treatment. Trop Biomed 2:30–33

Callaway E (2016) Dogs thwart effort to eradicate Guinea worm: epidemic in dogs complicates push to wipe out parasite. In: Nature News, Vol 529. http://www.nature.com/news/dogs-thwart-effort-to-eradicate-guinea-worm-1.19109

Rogers K (2016) Guinea worm disease. In: Encyclopedia Britannica. https://www.britannica.com/science/guinea-worm-disease. Accessed 30 Aug 2016

WHO (2016) Dracunculiasis – historical background. In: WHO http://www.who.int/dracunculiasis/background/en/ Accessed 28 Oct 2016

Chapter Ten

Editorial (1933) Amebic dysentery in Chicago. Am J Pub Health 23:1294–1295

Editorial (1934) Amebic dysentery in Chicago. Am J Pub Health 24:755–756

Ho AY, Lopez AS, Eberhart MG et al (2002) Outbreak of cyclosporiasis associated with imported raspberries, Philadelphia, Pennsylvania, 2000. Emerg Infect Dis 8:783–788

Istre GR, Dunlop TS, Gaspard GB et al (1984) Waterborne giardiasis at a mountain resort: evidence for acquired immunity. Am J Pub Health 74:602–604

Jarroll EL, Bingham AK, Meyer EA (1981) Effect of chlorine on Giardia lamblia cyst viability. Appl Environ Microbiol 41:483–487

Kwa BH, Moyad M, Pentella MA et al (1993) A nude mouse model as an in vivo infectivity assay for cryptosporidiosis. Water Sci Technol 27:65–68

MacKenzie WR, Hoxie NJ, Procter ME et al (1994) A massive outbreak in Milwaukee of cryptosporidium infection transmitted through the public water supply. N Eng J Med 331:161–167

O'hara SP, Chen XM (2011) The cell biology of cryptosporidium infection. Microbes Infect 13:721–730

Ralston KS, Petri WA (2011) Tissue destruction and invasion by *Entamoeba histolytica*. Trends Parasitol 27:254–263

Chapter Eleven

Cohen J (2011) What killed Charles Darwin? In: History 11 May 2011 http://www.history.com/news/what-killed-charles-darwin

Flegr J (2007) Effects of *Toxoplasma* on human behavior. Schizophr Bull 33:757–760

Jacobson R (2014) What happens when an amoeba "eats" your brain? In: Sci Am 18 July 2014 https://www.scientificamerican.com/article/what-happens-when-an-amoeba-eats-your-brain/

McAuliffe K (2012) How your cat is making you crazy. The Atlantic (March 2012)

McAuliffe K (2016) This is your brain on parasites. Houghton, Mifflin Harcourt, New York

Sugden K, Moffitt TE, Pinto L, et al (2016) Is *Toxoplasma gondii* infection related to brain and behavior impairments in humans? Evidence from a population-representative birth cohort. In: PLOS 10th Anniv. http://journals.plos.org/plosone/article?id=10.1371/journal.pone.0148435

Tenter AM, Heckeroth AR, Weiss LM (2000) *Toxoplasma gondii*: from animals to humans. Int J Parasitol 30:1217–1258

Chapter Twelve

Aponte JJ, Schellenberg D, Egan A et al (2009) Efficacy and safety of intermittent preventive treatment with sulfadoxine-pyrimethamine for malaria in African infants: a pooled analysis of six randomized, placebo-controlled trials. Lancet 374:1533–1542

Arama C, Troye-Blomberg M (2014) The path of malaria vaccine development: challenges and perspectives. J Int Med 275:456–466

Cisse B, Sokhna C, Boulanger D et al (2006) Seasonal intermittent preventive treatment with artesunate and sulfadoxine-pyrimethamine for prevention of malaria in Senegalese children: a randomized, placebo-controlled, double-blind trial. Lancet 367:659–667

Hsu E (2006) The history of *qing hao* in the Chinese *materia medica*. Trans Roy Soc Trop Med Hyg 100:505–508

Kayentao K, Kodio M, Newman R et al (2005) Comparison of intermittent preventive treatment with chemoprophylaxis for the prevention of malaria during pregnancy in Mali. J Infect Dis 191:109–116

Lee KS, Cox-Singh J, Singh B (2009) Morphological features and differential counts of *Plasmodium knowlesi* parasites in naturally acquired human infections. Malaria J 8:73

Miura K (2016) Progress and prospects for blood-stage malaria vaccines. Expert Rev Vaccines 15:765–781

Mordmuller B, Surat G, Lagler H et al (2017) Sterile protection against human malaria by chemoattenuated PfSPZ vaccine. Nature 542:445–449

Steverding D (2008) The history of African trypanosomiasis. Parasit Vect 1:3

Printed in the United States
By Bookmasters